Introduction to Chemistry
by
Russell J. Jeffers

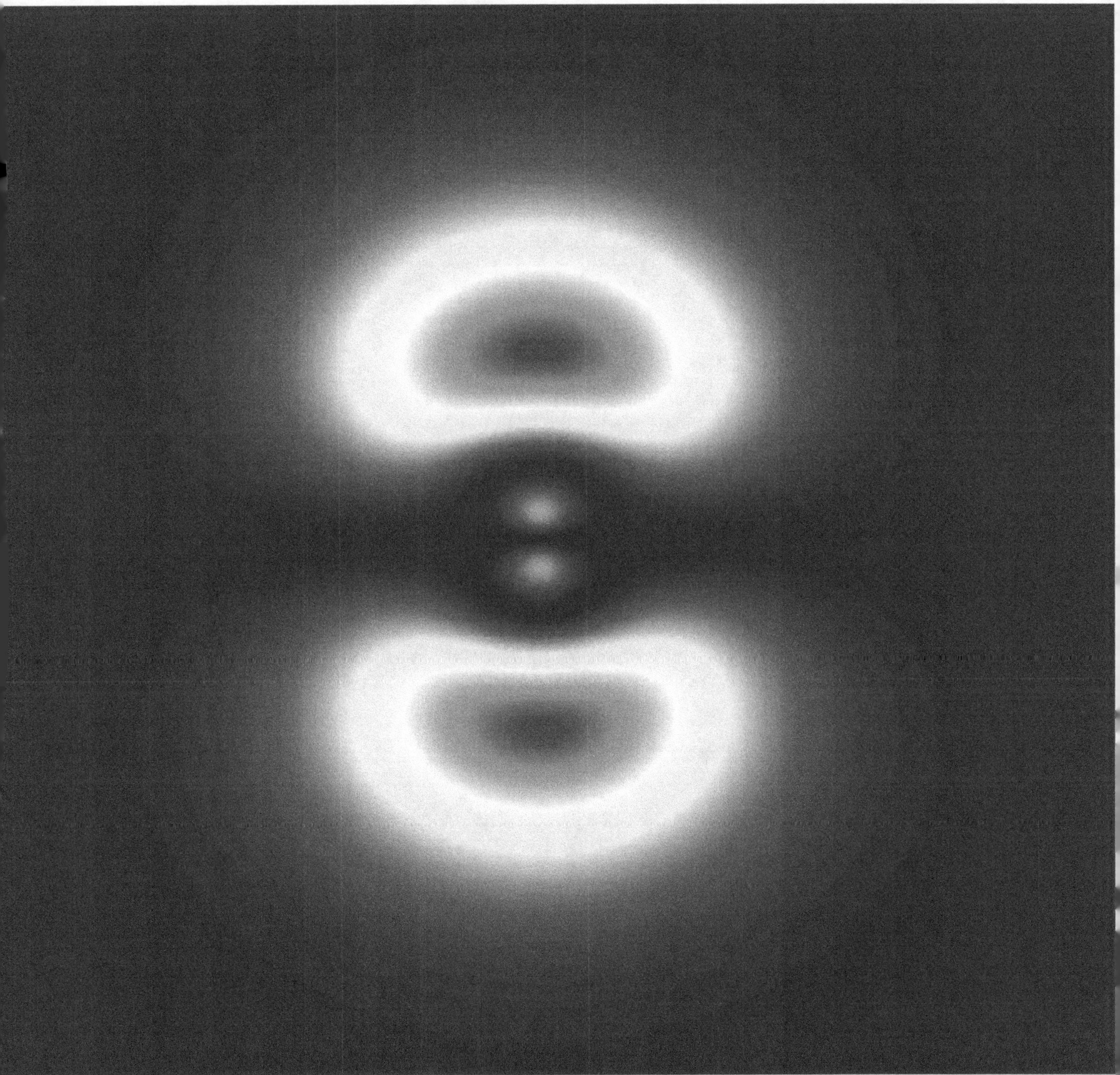

Introduction to Chemistry
A Series of Lectures

Russell J. Jeffers

Copyright © 2014 Russell J. Jeffers

This book and its on-line version are distributed under the terms of the Creative Commons Attribution-NonCommercial-ShareAlike 4.0 International license. A reference copy of this license may be found at http://creativecommons.org/licenses/by-nc-sa/4.0/legalcode

Noncommercial uses are thus permitted without any further permission from the copyright owner.

To

Claudia, Meyer, and Paul

Contents

Contents iv
Preface vi

Chapter 1 : The Atom 1
    The atomic theory 1
    Discovery of the electron 2
    J.J. Thomson experiments 3
    Discovery of the atomic nucleus 6
    Rutherford model of the atom 6

Chapter 2 : Early Quantum Physics, Wave-Particle Duality and the Bohr Model of the Atom 9
    Properties of light waves 9
    The photoelectric effect – Einstein's photon theory 13
    Photon momentum 15
    Particles as waves 16
    Failure of the classical description of the atom 17
    The Bohr model of the atom 17

Chapter 3 : Introduction to Quantum Mechanics 24
    Erwin Schrödinger 24
    Wave functions of a vibrating string 24
    Schrödinger equation 26
    Solutions to the Schrödinger equation for one-electron atoms 27
    s-orbitals: spherically symmetric wave functions 28
    Interpretation of the wave function 29
    Probability density functions 30
    Quantum numbers n, ℓ, m 32
    p orbitals for the single electron atom 33

Chapter 4 : Multi-Electron Atoms 40
    Introduction 40
    Electron spin 40
    Pauli exclusion principle 42
    Wave functions for multi-electron atoms 43
    Hartree theory 44
    Binding energies and the aufbau principle 47
    Periodic behavior and the periodic table 52

Chapter 5 : Chemical Bonds 61
    Covalent bonds 61
    Diatomic molecules 61
    Lewis structures 64
    Lewis structure examples 66
    Formal Charge 71

Electronegativity scale of the elements 73
Resonance 74
Exceptions to the Octet Rule 75
Ionic Bonds 79

Chapter 6 : Kinetic Theory of Gases 84
Ideal gas law, equilibrium and detailed balance 84
Kinetic theory 87
Mean free path 93
Boltzmann distribution 96

Chapter 7 : Internal Degrees of Freedom of Gases 99
Internal degrees of freedom 99
Energy levels of vibrational modes 102
Energy levels of rotational modes 106
Heat capacity of diatomic gases 108

Chapter 8 : Thermodynamics and Chemical Equilibrium 112
The laws of thermodynamics 112
What is entropy? 114
State variables 118
Gibbs energy and spontaneity 118
Spontaneous exothermic and endothermic reactions 120
Effect of temperature on spontaneity 122
Calculating standard entropy of reaction 123
Calculating standard enthalpy of reaction 124
Nature of chemical equilibrium 126
Equilibrium equation and rates of reaction 126
Chemical equilibrium and Gibbs energy 128
Temperature dependence of chemical equilibrium 130

Preface

This textbook is intended for the first college chemistry course, or an equivalent self-study course at that level. For example, this could be a useful companion to online courses. It is expected that students will have knowledge of calculus, introductory physics, and high school chemistry. I don't actually use very much calculus in the text, but understanding the definition of a derivative and integral are necessary. In college chemistry courses, there are problem assignments and lab assignments for which there are open source references available. I wanted to develop a set of lectures to describe the important chemical principals without many printed pages to solve example problems.

My intention was to give the key principles in each of the chapters. Derivations are provided when necessary to give insight. Otherwise results are simply stated and it is assumed that any diligent student could look up derivations themselves online. This helped keep the text light weight. Any students who feel that there is insufficient background can go to the internet to fill in details. The text provides the road signs.

The chapters map to very standard college lectures. In fact, different universities may cover the subjects in different orders. An example mapping of the chapters to a number of lectures is as follows. The italicized chapters are future chapters that I intend to write in my present optimistic state.

Chapter	Subject	Lecture
1	Atoms	1, 2
2	Early Quantum Physics	3, 4
3	Quantum Mechanics	5,6,7
4	Multi-Electron Atoms	8,9,10
5	Chemical Bonds	11,12
6	Kinetic Theory of Gases	13, 14
7	Internal Degrees of Freedom of Gases	16, 17
8	Chemical Thermodynamics and Reaction Rates	18, 19, 20
9	*Acids and Bases*	
10	*Redox Reactions*	
11	*Radioisotope research*	

Chapter 1 : The Atom

The atomic theory

The atomic hypothesis, first stated by John Dalton in 1805: *all substances consist of small particles of matter, of several different kinds, corresponding to the different elements.* Science textbooks often credit the Greek philosopher Democritus with first hypothesizing atoms; however, he provided no useful predictions to test his hypothesis, unlike Dalton. With regard to Democritus, the physicist James H. Jeans said memorably *"Given that a great number of thinkers are speculating as to the structure of matter, it is only in accordance with the laws of probability that some of them should arrive fairly near to the truth."*

The atomic hypothesis, having been experimentally confirmed for more than a century is now referred to as the atomic theory. A significant chemical law that influenced Dalton's thinking is called the *law of constant proportions*: *different samples of a chemical compound always contain elements in the same proportions by mass.* For example, the elements hydrogen and oxygen are always present in water (H_2O) in the proportion 1:8 for mass of hydrogen to mass of oxygen. This is consistent with Dalton's atomic theory.

Dalton made a prediction derived from the atomic theory, the *law of simple multiple proportions*. This was the first great success of atomic theory which was subsequently confirmed by experiment. The law of simple multiple proportions can be stated as follows: *If two elements form more than one compound between them, then the masses of one element which combine with the same mass of the other are in the ratio of small whole numbers.* For example, hydrogen and oxygen can combine to form water (H_2O) and hydrogen peroxide (H_2O_2). The mass ratio of hydrogen to oxygen is 1:8 for water and 1:16 for hydrogen peroxide. A quantity of hydrogen, say 100 g combines with 800 g of oxygen in water and 1600 g for hydrogen peroxide, so the ratio of oxygen in the two compounds is 1:2.

A second example of the law of simple multiple proportions is nitrogen combining with oxygen to form nitrous oxide (N_2O) and nitrogen dioxide (NO_2) in a ratio of 7:4 and 7:16 respectively. Thus, a quantity of nitrogen will combine with quantities of oxygen in a mass ratio of 1:4.

The atomic theory is the most important chemical theory. To quote the physicist Richard Feynman:

"If, in some cataclysm, all of scientific knowledge were to be destroyed, and only one sentence passed on to the next generation of creatures, what statement would contain the most information in the fewest words? I believe it is the atomic hypothesis (or the atomic, fact, or whatever you wish to call it) that all things are made of atoms – little particles that move around in perpetual motion, attracting

each other when they are a little distance apart, but repelling upon being squeezed into one another."

The atomic theory is a significant example of deductive reasoning in science. Analyses of chemical observations were used to deduce properties of objects that could not be observed directly. Another significant example of methods for deducing properties of things we cannot see is the discovery of the electron.

Discovery of the electron

At the time of the discovery of the electron in the early 20th century, physicists were still debating the reality of atoms. The critical experiments leading to the electron came out of research into *cathode rays*. In the late 19th and early 20th century, investigators had discovered peculiar "rays" in evacuated glass tubes, while investigating fluorescent gases, like the common fluorescent neon tube used for illuminated signs. It had been observed that an electric current was carried from the anode to the cathode, even when the glass tube was evacuated so completely that no gas fluorescence was observable.

Figure 1-1 illustrates an example of a so-called *cathode ray tube*. An electric potential between the cathode and the anode induces a current through gas in a glass tube. The *cathode* in this case is a metal electrode that carries an excess negative charge. The electrode that carries and excess positive charge is called the *anode*.

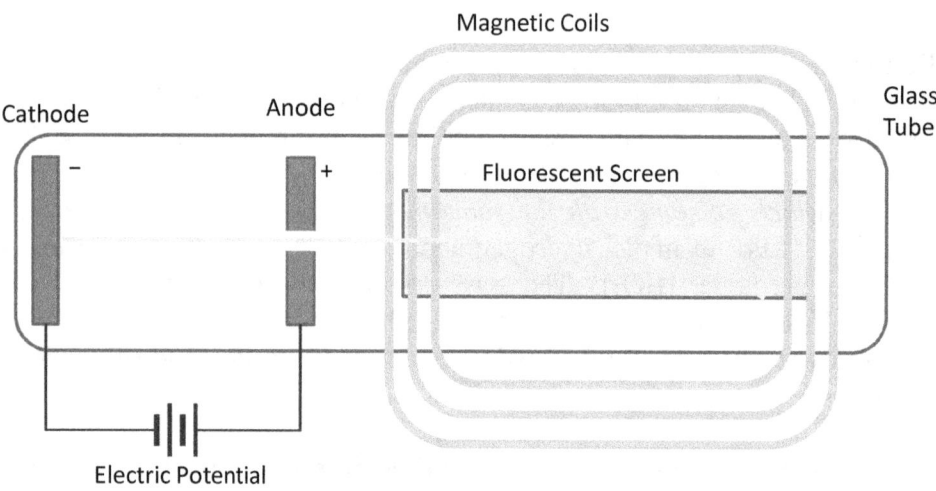

Figure 1-1: Diagram of a cathode ray tube illustrating the approach Jean B. Perrin used to demonstrate cathode rays are identical negatively charged particles.

When a cathode ray tube was constructed with a small hole in the anode, a beam of electrons accelerated from to the anode from the cathode and a narrow beam of electrons passes through the opening in the anode. A fluorescent screen is placed at the end of the tube, which consists of a sheet of paper or glass coated with a material which shines brightly when electrons impinge upon it. An example of a fluorescent material is zinc sulfide. Early investigators did not yet know this was a beam of electrons and referred to the phenomenon as cathode rays.

In 1895, Jean Perrin showed that this beam could be deflected by a magnetic field. A magnet's two poles were placed on either side of the beam of particles which creates magnetic flux lines at a right angle to the beam. The resulting force on a charged particle with charge q, moving with velocity of magnitude v, in a magnetic field of strength B is

$$F = qvB . \tag{1-1}$$

In the experimental setup by Perrin, the velocity and magnetic field are at right angles, so the direction of the force would be up or down depending upon the charge of the particles. The deflection of the cathode ray beam indicated that cathode rays were negatively charged particles. Had the beam been composed of particles with different charges the magnetic field would have separated the beam since each constituent would be deflected by a separate amount. Thus Perrin showed that cathode rays were not positively charged and were identical negatively charged particles.

J.J. Thomson experiments

Thomson was researching cathode rays himself, and had observed the cathode ray deflection due to a magnetic field. He was particularly impressed that the amount of deflection did not depend upon the gas in the cathode ray tube, nor did it depend upon the material that comprised the electrodes. He came to believe that they were charged particles rather than rays, and he also made estimates of the size of these particles based upon the following observation. The cathode ray beam traveling outside the tube lost one half the intensity when traveling a distance of 1 centimeter (cm), presumably by absorption due to air molecules. Thomson compared that distance traveled with an estimate of the mean free path of air molecules at atmospheric pressure which is 10^{-5} cm. He also observed that absorption of the particles (cathode rays) decreased with gas density, and did not depend upon the type of gas. From these observations he concluded that the size of these negatively charged particles are small compared with dimensions of molecules.

Thomson was able to measure the ratio of electron charge to mass, q/m using a cathode ray tube similar to Figure 1-2. The magnetic coils and electric plates allowed the deflection of the electron beam using either and electric field or a magnetic field. The calculation of the ratio of charge to mass is as follows. The force of a charge in an electric field is

$$\mathbf{F} = q\mathbf{E} \tag{1-2}$$

so the vertical displacement, y, of the charge moving between the electric plates is due to a force qE, applied for t seconds. The displacement due to the electric field-induced acceleration a, is

$$y = \frac{1}{2}at^2 = \frac{qE}{2m}t^2 \tag{1-3}$$

where the horizontal motion equal to the length of the plates, *L*, is calculated using the particle velocity, v,

$$L = vt \qquad (1\text{-}4)$$

so

$$y = \frac{qE}{2m}t^2 = \frac{qEL^2}{2mv^2} \qquad (1\text{-}5)$$

where *y* is calculated from the measured displacement of the beam at the fluorescent screen.

The next step is to apply a magnetic field using the magnetic coils which deflects the particle with an opposite force so that the magnetic and electric field-induced forces cancel. The equation for the total force on the particle is

$$\mathbf{F} = q\mathbf{E} + q\mathbf{v} \times \mathbf{B}. \qquad (1\text{-}6)$$

When the opposing forces exactly cancel, the magnitude of the force induced by the electric and magnetic fields are equal:

$$qE = qvB \qquad (1\text{-}7)$$

so

$$v = \frac{E}{B}. \qquad (1\text{-}8)$$

Substituting the particle velocity back into Equation (1-5) provides the desired ratio

$$\frac{q}{m} = \frac{2yE}{B^2L^2} \qquad (1\text{-}9)$$

So Thomson was able to compute the velocity of these particles and their ratio of charge to mass. The charge to mass ratio was very high compared with other particles such as ionized molecules, which suggested a small mass. Together with the observation that the charge to mass ratio was constant for different experiments, and using his earlier inference that the particles are significantly smaller than air molecules, he deduced that these particles were "corpuscles" much smaller than atoms.

Thomson went on to interpreting the structure of the atom, as George Gamow describes it:

> *"Thomson visualized the atom as being formed by some positively charged substance distributed more or less uniformly through the entire body, with negatively charged electrons imbedded in it as are raisins inside a round loaf of raisin bread. The electrons are attracted to the center of the positive charge distribution*

and repelled by one another according to Coulomb's law of electric interactions, and the normal state of the atom is attained when these two opposing sets of forces are in equilibrium."

The Thomson model of the atom was not to be described as round raisin bread, but was to become known as the plum pudding model of the atom, named after a British dessert. This model of the atom implied that ionization was like splitting up an atomic particle into the electron and remaining ion. This is essentially correct, but was difficult for physicists of the time to accept. Today, Thomson is credited for the discovery of the electron.

In 1909, the American physicist Robert Millikan measured the charge of the electron using the ingenious *oil drop experiment*. Oil drops, produced by a fine atomizer, acquired a negative charge using ionizing radiation. In the experiment, these drops were levitated in air using an electric field to cancel the force of gravity. The mass of the droplet was computed by measuring the diameter of the oil droplet, and knowing the droplet mass and the electric field strength, the charge of the droplet could be calculated. The electric charge was shown to be quantized, taking on discrete values. Millikan was able to accurately calculate the electron charge. Based upon the measurement of the electron's charge, the electron mass was then known to be 1/1836 times the mass of the hydrogen atom.

Later, beta rays were shown to have the same charge to mass ratio as the electron; however, as will be discussed in the next section, beta rays emanate from the *nucleus* of the atom.

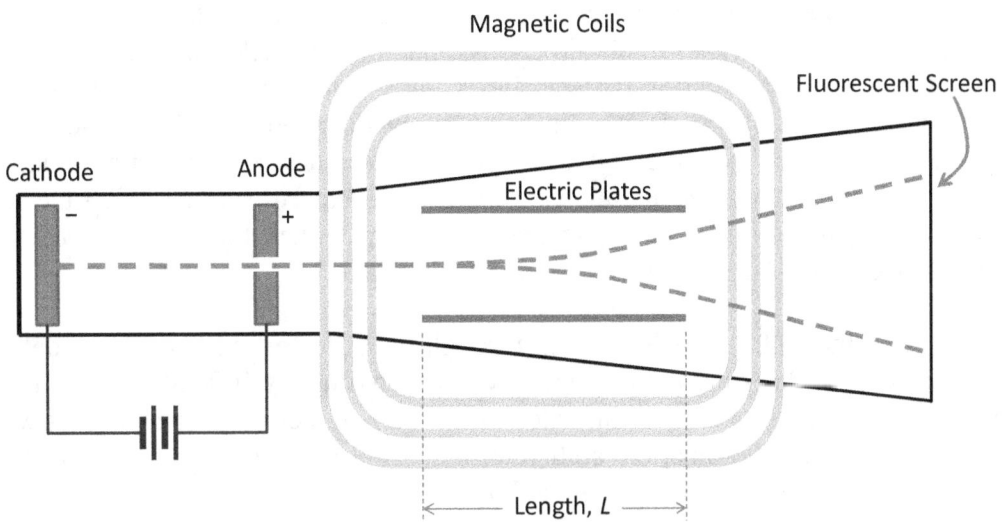

Figure 1-2: Diagram of a cathode ray tube illustrating the approach J.J. Thomson used to measure the charge to mass ratio of the electron.

Discovery of the atomic nucleus

Ernest Rutherford conducted studies on radioactivity under the supervision of J.J. Thomson and discovered two distinct types of radioactivity, which he named *alpha* and *beta radiation* after the first two letters of the Greek alphabet. Using magnetic fields to deflect a collimated beam of radioactive particles, it was shown that alpha particles are positively charged, and beta particles are negatively charged. Beta particles were shown to be very similar to cathode rays and were eventually shown to be electrons. A third type of radioactivity was discovered and following Rutherford's lead, named *gamma radiation*. Gamma radiation was eventually shown to be high energy electromagnetic radiation, much like x-rays.

Rutherford started his own research lab and showed that ionized helium atoms comprised alpha radiation. He designed an experiment, in which a radioactive source emitted alpha radiation through a foil barrier into a glass vessel. Over time the glass vessel accumulated gaseous helium which was shown to be present spectroscopically. Rutherford concluded correctly that alpha particles were helium atoms with their two electrons removed; however, the Thomson model of the atom implied that the twice ionized helium atom should be nearly the size of the non-ionized atom. Rutherford was to conduct experiments which demonstrated that most of the mass of an atom was contained in a small volume compared with the total volume of the atom.

Rutherford model of the atom

In the years between 1909 and 1913, Rutherford conducted a series of experiments together with Hans Geiger and Ernest Marsden which involved the scattering of alpha particles. One of the most significant experiments conducted by Geiger and Marsden used a thin gold foil to scatter alpha particles. The experiment is illustrated in Figure 1-3.

A radioactive source, in this case radium, is placed behind a lead block. The radium source emits alpha particles in all directions, and alpha particles are quickly absorbed by lead. A narrow hole bored in the lead block allows the formation of a narrow beam of alpha particles. The beam of alpha particles can be detected by a zinc sulfide fluorescent screen. Rutherford had observed the alpha particles are scattered, thus widening the beam, by many materials including air. The critical experiment quantified the scattering using a very thin foil of gold. Geiger and Marsden looked for and observed scattering events at extremely large angles many greater than 90^0.

Most of the alpha particles passed through the gold foil without being deflected. A gold foil of 4000 Angstroms thick, which was approximately 1000 atomic layers, deflected one particle in 100,000. As illustrated in Figure 1-3, these scattering events were often at large angles. At the time, the ability for an atom to effectively scatter alpha particles was not predicted by the Thomson model of the atom. Rutherford said:

"It was quite the most incredible event that has ever happened to me in my life. It was almost as incredible as if you had fired a 15 inch shell at a piece tissue paper and it came back and hit you."

The alpha particles are very energetic, with the kinetic energy equivalent to acceleration through millions of volts. They had approximately the mass of a helium atom, so to be backscattered, even rarely, by the gold foil, the alpha particles have to encounter something charged and equally massive (or larger). Furthermore, the object has to be small in diameter. Rutherford explained the experimental effects with a new model of the atom that assumed close encounters between alpha particles and the electrical field of a small volume charge, or *nucleus* which is what it came to be called later. The observation that one alpha particle in 100,000 (10^5) was scattered by a layer of gold atoms 1000 (10^3) thick led Rutherford to estimate volume of the nucleus using the following reasoning. A foil of gold 1 atom thick would scatter 1 alpha particle in 10^8. This leads to the calculation that the cross sectional area of the nucleus is a factor 10^8 smaller than the cross section of an atom. The diameter of the nucleus would then be approximately 10^4 smaller than the diameter of an atom. It is often said that an assumption for the Rutherford model is the atom is mostly empty space.

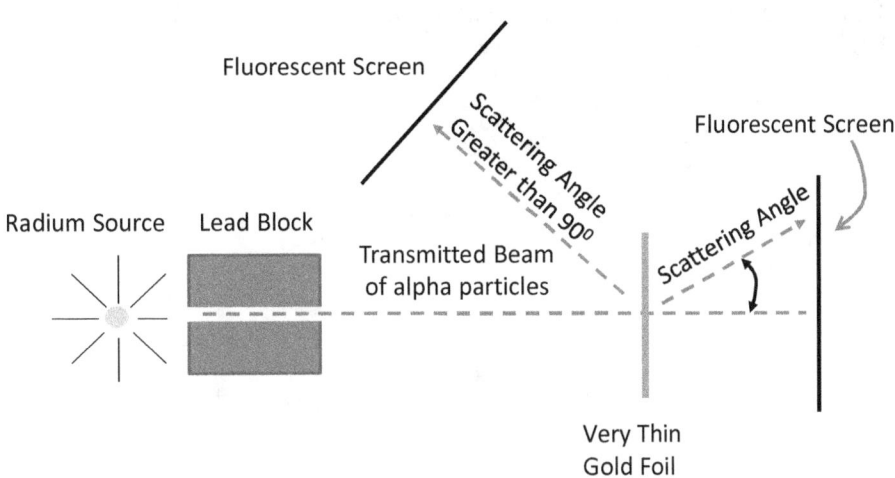

Figure 1-3: Diagram of the gold foil scattering experiment conducted by Ernest Rutherford, Hans Geiger and Ernest Marsden.

The Coulomb force law for the attraction of the electron to the positive nucleus is an *inverse square law*, meaning that the attractive force is proportional to the inverse of the electron-nucleus distance squared. The gravitational force law is also an inverse square law. It is because of the similar force law that Rutherford visualized the atom as having the shape and dynamics of a small solar system with the electron orbiting the nucleus. However, the next chapter will describe how such a model of the atom is unstable, and the modern model of the atom will be introduced.

References

Davis, E.A and Falconer, I.J., *J.J. Thomson and the Discovery of the Electron*, CRC Press, 1997.

Feynman, R., Leighton, R.B., Sands, M., *Lectures on Physics*, Adeson-Wesley Publishing Company, 1966.

Gamow, G., *Thirty Years That Shook Physics*, Dover Publications, Inc., 1966.

Jeans, J.H., *The Dynamical Theory of Gases*, 2nd Edition, Cambridge at the University Press, 1916.

Pauling, L., *College Chemistry*, W.H. Freeman and Company, 1957.

Halliday, D. and Resnick, R., *Fundamentals of Physics*, 2nd Ed., John Wiley & Sons, 1981.

Chapter 2 : Early Quantum Physics, Wave-Particle Duality and the Bohr Model of the Atom

Introduction

In this chapter, modern quantum physics will be introduced. In the early 1900s as the concept of the atom was being accepted, new discoveries were made about the properties of light. The *wave-particle duality* of light will be discussed. In many experiments and observations, light is observed to have wave-like properties. In other experiments described here, light consists of particles referred to as *photons*. Furthermore, matter as well as light has wave-like and particle-like properties. The modern mathematical description of matter which accounts for the wave-particle duality of matter is called *quantum mechanics*.

Properties of light waves

Visible *light* is composed of *electromagnetic waves*. In fact, there are many forms of electromagnetic waves that are not visible light. An illustration of the many forms of electromagnet waves are shown in Figure 2-2. Together, they are referred to as the *electromagnetic spectrum*. Light and electromagnetic waves are considered synonymous and the terms are hereafter used interchangeably.

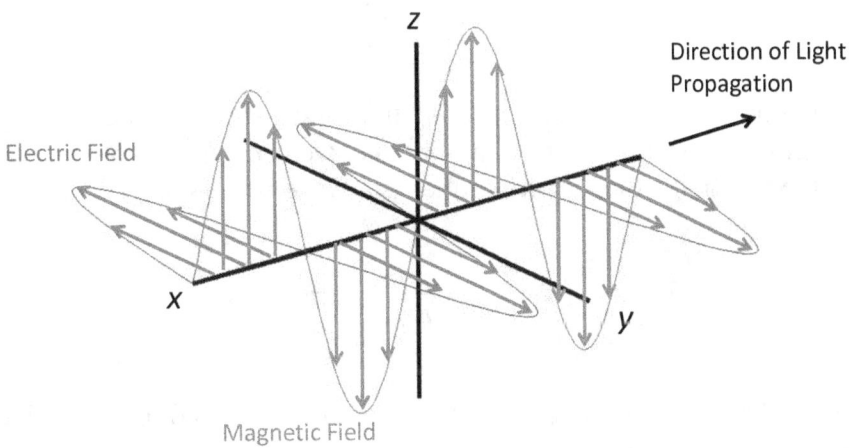

Figure 2-1: Illustration of propagating electromagnetic wave. The electric field in blue is always at right angles to the magnetic field in red. Both field directions are always at right angles to the direction of propagation.

The illustration of an electromagnetic wave in Figure 2-1 illuminates some of the properties of electromagnetic (EM) waves. A periodic *electric field* and periodic *magnetic field* comprise the EM wave. The fields are propagating in space, so that the fields vary periodically in time as well as space. The illustration in Figure 2-1 shows the spatial distribution of the fields at a single instant in time. The electric field and magnetic field are each a vector quantity, which means that the fields always have a direction. The pointing directions of the electric field and magnetic field are always at 90° angles to each other. Vectors which are at 90° angles to each other are said to be *orthogonal*.

The magnetic and electric field directions are always orthogonal to each other, and they are always orthogonal to the direction of propagation of the wave. In Figure 2-1, the light wave is shown plotted in Cartesian coordinates x, y and z. Note that the x, y and z axes are orthogonal to each other. In the figure, the electric field direction is always pointing in the positive or negative y-direction. The magnetic field is always pointing in the positive or negative z-direction. The light wave is propagating in the negative x-direction.

Wavelength and frequency

Light waves have a spatial periodic structure as shown in Figure 2-1. The periodicity in space is described by the *wavelength* (λ). The wavelength has units of length (m) and is the length of one cycle of the periodic function. The periodicity of light waves in time is described by the *frequency* (ν). The frequency has units of inverse time (s^{-1}). Sine and cosine periodic functions comprise the mathematical representation of electromagnetic waves. In chemistry, we are interested in the electric field component of a light wave since that is the component that interacts with electric charges. The electric field of an electromagnetic wave in the direction of the y-axis and propagating in the negative x-direction is

$$E = E_y \sin\left(\frac{2\pi x}{\lambda} + 2\pi \nu t\right) \quad (2\text{-}1)$$

where t is time and x is the distance along the x-axis. If we choose to analyze the periodic structure in space (x) at a given instant in time, we can conveniently define the time to be $t = 0$, so the function which describes the light wave is

$$E = E_y \sin\left(\frac{2\pi x}{\lambda}\right) \quad (2\text{-}2)$$

where it is observed that the electric field completes one complete periodic cycle in the distance between $x = 0$ and $x = \lambda$.

To analyze the periodic form of the light wave in time alone, we observe the electric field at a single point in space. Conveniently, setting $x = 0$, we see that the form of the wave function is

$$E = E_y \sin(2\pi v t) \tag{2-3}$$

where it is observed that the electric field will complete a single periodic cycle, or *period* in the time from $t = 0$ to $t = 1/v$. It is clear that the period of one cycle is $1/v$, and v has units of inverse time (s^{-1}). It is typically said that the *frequency represents the number of cycles per second*. The units of second^{-1} (s^{-1}) is typically referred to as the units of Hertz, abbreviated (Hz) named after Heinrich Hertz, the German physicist credited with experimental verification of the electromagnetic wave nature of light. Often, in speaking, there is confusion about whether Hz is singular or plural as in "one Hert" compared with "two Hertz". The unit of Hz is both singular and plural.

The mathematical formulation of the electromagnetic wave in Equation (2-1) does not give the speed of propagation unless the relationship between λ and v is given. For light waves, the speed of wave propagation is given by

$$c = \lambda v \tag{2-4}$$

where the *speed of light* is always represented by the variable c from the Latin celeritas, meaning speed or swiftness. Note that the wavelength and frequency are inversely related, so that given the wavelength, the frequency can be calculated

$$v = \frac{c}{\lambda}. \tag{2-5}$$

The speed of light in a vacuum is a constant for all wavelengths and all conditions and has the value

$$c = 2.998 \times 10^8 \text{ m/s}. \tag{2-6}$$

Propagation of electromagnetic waves through materials can slow the speed of light to a value less than c, but the maximum propagation speed, c, is achieved in a vacuum.

Intensity, constructive and destructive interference
An important physical property of waves is the *intensity*, the power in a unit of area. For visible light, the intensity is what is usually thought of as brightness. The formula for the intensity, I, of an electromagnetic wave in a vacuum is

$$I = \frac{1}{2} c \varepsilon_0 |E|^2. \tag{2-7}$$

where ε_0 is an electrical constant called the permittivity constant and I has units of watts per square meter (W/m^2). A consequence of this is the intensity is necessarily positive since it is proportional to the square magnitude of the electric field. Furthermore, intensity does not depend upon the sign or direction of the electric field.

An important property of waves generally, which will come up in following sections, is *interference*. Two waves can be added together and the outcome depends upon the sum of the electric fields of the waves. Since the electric fields can have positive and negative signs, summing the waves can have complicated interference effects. For example, suppose the electromagnetic wave in Figure 2-1 was added to a wave with equal electric field, but opposite direction. The sum of the light waves would be zero. Such an occurrence is referred to as *destructive interference* because the overall field intensity is reduced to zero. Alternatively, suppose the electromagnetic wave in Figure 2-1 was added to a wave which has identical electric field and direction. The electric field would increase by 2, and the intensity would increase by 4. This is known as *constructive interference* because the overall intensity is increased by the sum of the fields.

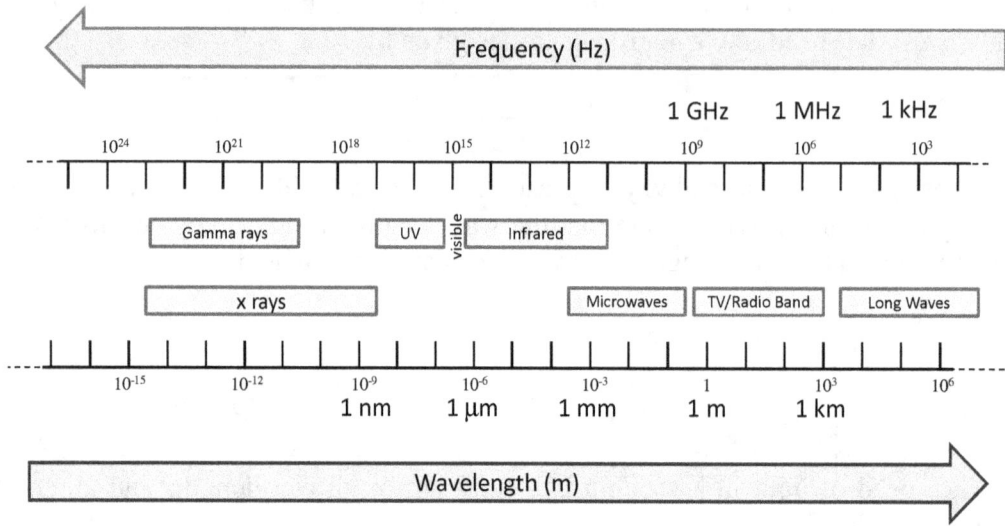

Figure 2-2: Illustration of electromagnetic spectrum. Several interesting categories of light waves are shown with increasing wavelength from left to right. Wave frequency decreases from left to right since it is inversely proportional to wavelength.

The electromagnetic spectrum

As discussed above, all propagating waves are characterized by a wavelength and a wave frequency. Examples of waves in the electromagnetic spectrum are shown in Figure 2-2. The wavelength of the categories of light waves increases from left to right in the figure. Since the frequency is inversely proportional to the wavelength, the frequency decreases from left to right in the figure. Interesting categories of light include:

visible spectrum – the familiar colors of the rainbow (to the human eye) which have wavelengths between 390 nm (violet) to 750 nm (red).

UV spectrum – UV stands for *ultra-violet*, or beyond violet, and has wavelengths longer than visible light between 10 nm to 390 nm. The sun produces significant intensity in this region, not visible to human eye but can cause sun burn.

infrared (IR) spectrum – the IR spectrum has wavelengths longer than visible light, between 750 nm and 300 μm. Radiated heat from a hot stove consists of IR waves.

X-rays – Created by accelerating electric charges, usually electrons. This category covers wavelengths shorter than the UV spectrum from 10 nm and shorter, and it overlaps the gamma ray spectrum.

gamma rays – largely overlap the X-ray spectrum. Differentiated from X-rays by the source. Gamma rays are emitted from atomic nuclei, while X-rays are emitted by accelerating charges interacting with matter.

microwave spectrum – longer than IR wavelengths, between 1mm and 1m. This spectrum is used for communications, radar and the familiar microwave oven.

longer waves – beyond the microwave spectrum are electromagnetic waves longer than 1m. This includes communications bands such as television, AM and FM radio.

The photoelectric effect – Einstein's photon theory

When a metal surface is illuminated with particular wavelengths of light, electrons are emitted from the surface (Figure 2-3). This is known as the *photoelectric effect*. It was observed experimentally that different metals ejected electrons in similar manner; however, not all wavelengths of light are equal in ejecting electrons. Sometimes red light does not eject electrons from a metal's surface, and increasing the intensity of the light does not cause electrons to be ejected. This intensity independence is surprising because increasing light intensity provides greater energy per unit area.

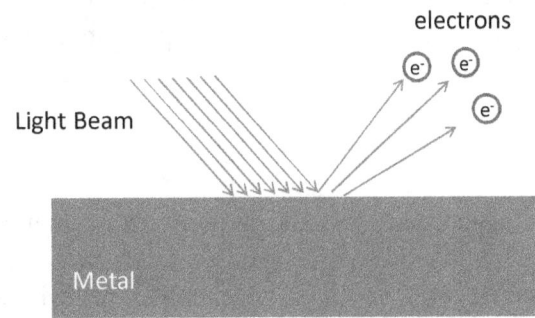

Figure 2-3: Illustration of the photoelectric effect. The illumination of metal surfaces with light of certain wavelengths ejects electrons from the surface of the metal.

It was observed experimentally, that in cases if light of one wavelength did not eject electrons, then illuminating with shorter wavelengths of light induced the photoelectric effect. It is worth noting, that decreasing wavelength is equivalent to increasing frequency. This phenomenon of the photoelectric effect induced above a particular frequency is known as the *threshold effect*. Furthermore, the kinetic energy of the ejected electrons is related to light frequency. At the threshold frequency, the ejected electrons barely have any speed, which means they have very low kinetic energy. Increasing the light intensity at the threshold ejects more electrons, but does not increase the kinetic energy (*i.e.* speed) of these ejected electrons. As higher frequencies are used the kinetic energy (speed) of the ejected electrons increase. This is illustrated in Figure 2-4 for two metals: sodium and iron. Below the threshold frequency, the photoelectric effect is not induced. As the frequency is increased the electron kinetic energy is increased linearly with the light frequency.

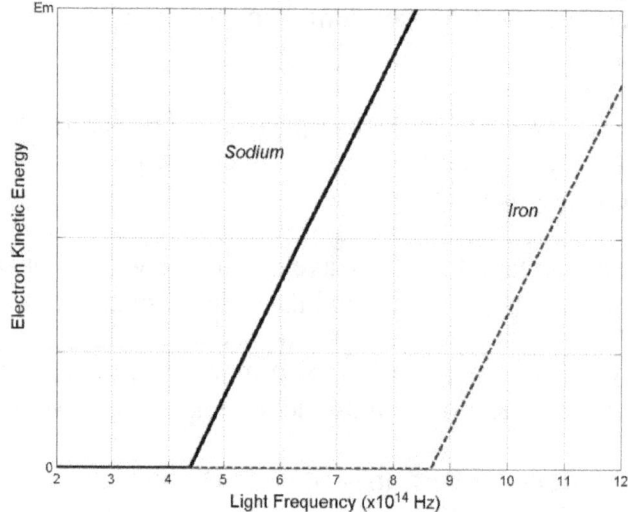

Figure 2-4: Electron kinetic energy for electrons ejected from a metal surface due to the photoelectric effect. Each metal has its own threshold light frequency where the photoelectric effect is induced and below which no electrons are ejected. Iron and sodium are shown. Note the sodium threshold frequency is in the visible region (λ=684 nm) while the iron threshold is in the UV region (λ=346 nm).

Photons

In 1905, Albert Einstein published an article in which he postulated a relationship between light energy and Planck's constant. His hypothesis was that light energy was carried in discrete, or quantum units of energy proportional to light frequency and the energy E for these light quanta is

$$E = h\nu \tag{2-8}$$

where h is Planck's constant and ν is the light wave frequency. Since this amounts to describing light as consisting of small particles, each carrying energy, the light quanta were later given the name *photons*. The name is meant to be analogous to other fundamental particles like the electron and proton.

Einstein's theory of photons was initially controversial but was ultimately verified by the photoelectric effect. Using Equation (2-8), experimental results like those illustrated in Figure 2-4 can be derived. There is some work required to extract an electron from the metal surface, and the quantity of the work is referred to as the *work function*, ϕ. Work has units of energy, and the work function can be viewed as the energy required to overcome the metal binding or "holding" the electron. The threshold frequency at which electrons are ejected with zero kinetic energy is when the photon energy is equal to the work function. Thus the kinetic energy of the electrons ejected from a metal's surface is equal to the energy contained in one photon minus the energy required to extract the electron from the metal surface

$$K = h\nu - \phi \tag{2-9}$$

where K is the electron kinetic energy. It is clear from Equation (2-9) that the photon theory predicts three aspects of the photoelectric effect: the threshold effect, linear proportionality between electron kinetic energy and light frequency, and kinetic energy independence of light intensity. It is also assumed that one photon is absorbed by one electron. It is not assumed that there are complicated sequences such as an electron absorbing one photon, then a second photon and so one. These predictions were later experimentally verified. The exact proportionality of K to ν, similar to that illustrated in Figure 2-4, was measured by Robert Millikan and published in 1914.

Photon momentum

The photoelectric effect provided the initial experimental verification of the photon theory of light, but the photoelectric effect is not the only evidence of photons. In addition to photons having discrete energy, Einstein derived the momentum of a photon. The equation for photon momentum is

$$p = \frac{h}{\lambda} = \frac{h\nu}{c} \tag{2-10}$$

in which λ is the wavelength of light. The American physicist Arthur Compton conducted an elegant experiment in 1923 demonstrating the reality of photons and verifying the equations for photon energy and momentum.

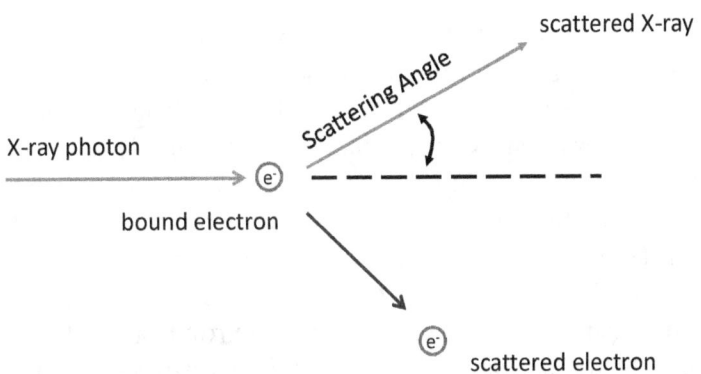

Figure 2-5: Diagram of the Compton Effect. The X-ray photon collides with an electron and the electron recoils in the same manner as colliding billiard balls. The energy transferred to the scattered electron reduces the scattered X-ray frequency. Energy and momentum are conserved.

The Compton Effect

Arthur Compton was born in Wooster, Ohio. His father was dean of the College of Wooster which Arthur attended prior to his successful career as a physicist including the 1927 Nobel Prize. He was awarded the Nobel Prize in physics for demonstrating what has become known as the *Compton effect*.

Compton wanted to conduct an experiment in which photons are scattered by electrons. He chose to illuminate a target with X-rays which have extremely high photon energy due to their high frequency. The binding energy of outer electrons in molecules is negligible because it is so small compared to the X-ray's photon energy. The X-ray photon collides with the electron causing the photon to lose energy and momentum to the recoiling electron (Figure 2-5). By analyzing the X-ray scattering in the same manner that billiard balls collide and scatter, Compton predicted the scattered X-ray photon frequency using Equation (2-10). He was able to relate the X-ray scattering angle to the change in photon frequency using conservation of energy and momentum.

The Compton effect and the photoelectric effect validate the photon energy and momentum equations. Light is said to exhibit *wave-particle duality*. It can have particle-like properties in some experiments, and wave-like properties (e.g. interference) in other experiments. Light is treated as either a wave or a particle depending upon the experiment.

Particles as waves

Louis de Broglie postulated in his Doctoral thesis in 1925 that particles, such as electrons, nuclei and atoms could have wave-like properties. This was analogous to the concept of light having particle-like properties. Inspired by Einstein's relation between wavelength and momentum for photons, the wavelength for matter waves was determined by de Broglie to be inversely related to the particle's momentum and the proportionality constant is Plank's constant.

$$\lambda = \frac{h}{p} = \frac{h}{mv} \qquad (2\text{-}11)$$

Electron beams accelerated through several kilovolts of electric potential can be shown to have wavelengths of the same order of magnitude as X-ray wavelengths. In order to observe the wave properties of matter, the wavelength is required to be on the scale of the equipment interacting with the waves. For example, a *diffraction grating* used to analyze light spectra has grating spacing on the order of light wavelengths. Electron beams can be accelerated to a momentum so the wavelength is 10^{-10} m which is on the order of crystal layer spacing in a crystalline solid (Figure 2-6). The beam of fast electrons can be viewed as waves, and the waves reflect off multiple layers and produce constructive and destructive interference. An electron beam detector will detect the interference. The detector output will be bright for constructive interference and dark for destructive interference. At different angles of incidence, the geometry will result in a series of bright and dark fringes called an interference pattern.

This type of electron scattering and diffraction experiment was carried out independently by British physicist George Thomson and by American physicists G. Davisson and L.H. Germer. Electron diffraction and interference proved the reality of the wave-like nature of matter. Matter, like light, is said to exhibit wave-particle duality. In some experiments matter behaves like particles, like electrons. J.J. Thomson's experiments, for example, showed the particle-like properties of electrons. In other experiments, like electron diffraction, matter behaves as waves. It is interesting that George Thomson who proved the wave-like properties of electrons is the son of J.J. Thompson who is credited with proving the electron is a particle.

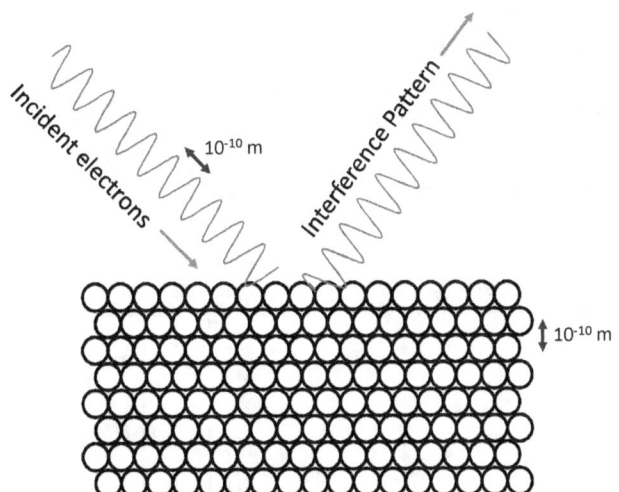

Figure 2-6: Illustration of electron diffraction by a crystalline solid. The beam of fast electrons has wavelength on the order of the spacing of the successive layers in the crystal structure. The waves reflect off multiple layers and produce constructive and destructive interference, resulting in a series of bright and dark fringes depending upon the angle of incidence.

Failure of the classical description of the atom

The Coulomb force law describes the attractive force between the negatively charged electrons and the positively charged nucleus. It is an inverse square law as is the gravitational law of attraction, so the electrons can move around the nucleus in elliptical orbits as the planets move around the sun. There is a great difference in that the planets and sun are electrically neutral. The electron revolves around the nucleus, and it is well established in classical electromagnetics that oscillating electric charges radiate electromagnetic waves (including light waves). The electron orbits would be like a very small antenna radiating at very high frequency. The radiated electromagnetic waves carry away all of the electron's energy in a time less than 10^{-8} second, and the electron collapses into the nucleus. Thus the Rutherford model of the atom is unstable.

The Bohr model of the atom

The Danish physicist Niels Bohr was inspired by the experimentally proven Rutherford model of the atom; however, he was intrigued by its theoretical impossibility. Bohr assumed that the mechanical energy in an atom must be quantized into discrete energy levels analogous to light energy existing in discrete quanta. This assumes that Newtonian mechanics does not hold for very small scales, like the atomic scale. Bohr focused on modeling the hydrogen atom because it was the simplest element: one electron and one proton.

Bohr formed a hypothesis that the hydrogen atom has discrete energy levels. The lowest energy state is called the ground state, and the atom has higher energy states in discrete intervals. Bohr's first assumption was *the ground state of the atom is stable*. He used the *radiation spectrum* of the hydrogen atom to provide insight into the energy levels. If a gas like hydrogen is raised to a very high temperature, the atoms will be raised from the ground state to higher energy states brought about by energetic thermal collisions between themselves. When the atomic energy level drops from a higher energy to a lower energy it releases a photon which carries away the difference in energy. Another method for *exciting* atoms to an energy state higher than the ground state, or *excited state*, is to run an electric current through a gas. Electrons accelerated in the electric field collide with atoms, and excite them to higher energy state. This type of device, known as a *gas discharge tube*, is very common in lighted signs. The familiar neon sign uses this method to excite neon atoms which produce a visible radiation spectrum.

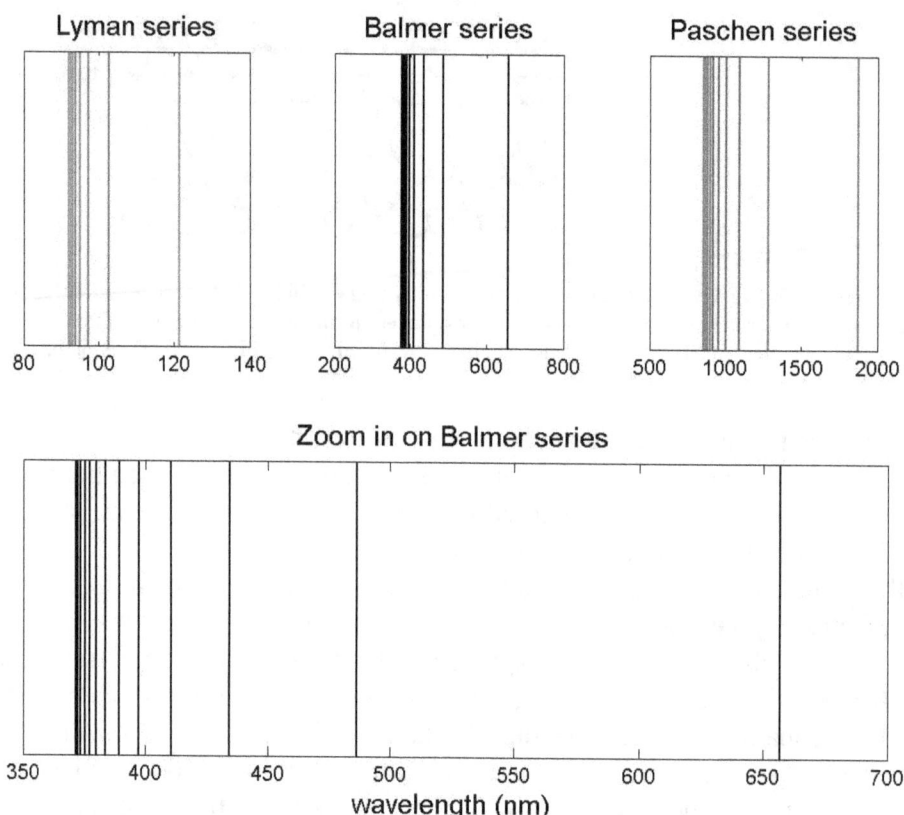

Figure 2-7: Illustration of the hydrogen line spectra plotted against the spectral line wavelength in nanometers (nm). The lines are shown as measured at various wavelengths which include groups of lines in the visible (Balmer series), ultraviolet (Lyman series) and infrared region (Paschen series).

Bohr relied on a mathematical model of the hydrogen spectrum developed in 1885 by Swiss schoolteacher J. J. Balmer who discovered that hydrogen excited using a gas discharge tube had a regular spectrum of discrete wavelengths, which are called *atomic line spectra*. These line spectra are illustrated in Figure 2-7 where lines are grouped into series of lines named for the discoverer of the line series. The series discovered by Balmer is in the visible spectrum. Balmer's mathematical formula calculated the frequency of the mth visible lines

$$v_m = R\left(\frac{1}{4} - \frac{1}{m^2}\right) \sec^{-1} \qquad (2\text{-}12)$$

where the constant R is called the *Rydberg constant*

$$R = 3.289 \times 10^{15} \text{ sec}^{-1} \tag{2-13}$$

and the wavelength of the mth visible line is

$$\lambda_m = \frac{c}{v_m} = \frac{c}{R}\left(\frac{1}{4} - \frac{1}{m^2}\right)^{-1} \text{ meter} \tag{2-14}$$

where c is the speed of light.

Bohr's hypothesis is that the mechanical energy of the atom is quantized so the photon energy carried away is equal to the difference between energy levels. Once can visualize an electron transitioning from a higher energy level to a lower energy level. In order to accomplish this transition, the electron would have to give up energy equal to the difference in the two energy levels. The photon energy accounts for the energy difference. For the Balmer series, the photon energy is

$$h v_m = hR\left(\frac{1}{4} - \frac{1}{m^2}\right) = hR\left(\frac{1}{2^2} - \frac{1}{m^2}\right) \tag{2-15}$$

A general series can be written out for a transition from a higher m^{th} energy level to a lower n^{th} energy level as

$$h v_{mn} = hR\left(\frac{1}{n^2} - \frac{1}{m^2}\right) = -hR\left(\frac{1}{m^2} - \frac{1}{n^2}\right) \tag{2-16}$$

where the Bohr energy levels are defined as

$$E_n = -\frac{hR}{n^2} \tag{2-17}$$

Representing the atomic energy levels as negative in value is consistent with a *binding energy* since the electron is bound to the nucleus. The zero energy would be when the particles are an infinite distance from each other. The physical significance of the negative binding energy is positive energy must be put into the atom in order to ionize it.

The more general formula (Equation (2-15)) for hydrogen line spectra derived by Bohr implies line families other than the Balmer series in the visible spectrum. Other series, Lyman and Paschen, are named after their discoverers. The Lyman series which is in the UV spectrum is due to electron transitions from higher energy levels to the hydrogen ground state. The Paschen series, in the infrared spectrum, is due to electron transitioning to the n=3 energy level from higher energy levels.

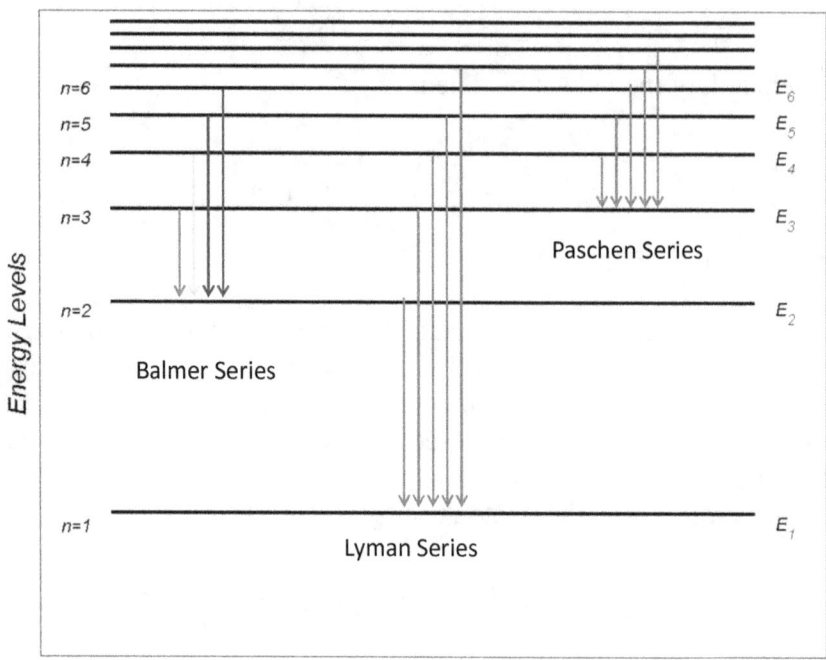

Figure 2-8: Diagram of Balmer, Lyman and Paschen Series of hydrogen spectra. The spectra are due to electrons transitioning from higher energy levels to lower, and the wavelength of the various spectral lines is calculated from the difference in energy.

Calculating the Bohr radius

The energy level interpretation was a reformulation of the Balmer formula, but the reason the Bohr model was so important was because Bohr introduced the idea of quantizing the mechanical energy of the atom in order to derive the Balmer formula. The derivation follows below.

The Coulomb force between the positively charged nucleus of charge +e and the electron of charge –e is

$$F_{Coulomb} = -\frac{e^2}{r^2} \qquad (2\text{-}18)$$

where r is the radial distance between the electron and the nucleus. For an orbit to be stable, the radial Coulomb force must be balanced by centrifugal force. The first approximation is to treat the much heavier nucleus as un-moving and the radial distance between the two particles is the orbital radius as well, then the centrifugal force is

$$F_{centrifugal} = +\frac{mv^2}{r} \qquad (2\text{-}19)$$

where v is the electron velocity and m is the electron mass. The balancing of the two forces produces the following equation

$$F_{centrifugal} + F_{Coulomb} = +\frac{mv^2}{r} - \frac{e}{r^2} = 0 \qquad (2\text{-}20)$$

which allows the calculation of the electron velocity v along a circle of radius r

$$v = \frac{e}{\sqrt{mr}} \qquad (2\text{-}21)$$

At this point it should be realized that classical physics allows the electron to have any value of v, but if the atomic energy is to be quantized, only electron velocities with atomic energy E_n in Equation (2-17) are allowed. Now the quantized Bohr orbits will be derived, and it is important to realize that the Coulomb force law is assumed to hold. Bohr's assumption was that atoms did not obey Newtonian mechanics, but classical electromagnetics still held at atomic scales.

The quantization rule that Bohr applied next was that *the total action of the electron is integer multiples of Plank's constant, h.* The action has units of mass × velocity × distance. The action of the electron in a closed circular orbit which must be quantized is

$$m \times v \times 2\pi r = nh \qquad (2\text{-}22)$$

so the nth radial orbit r_n moving with velocity v_n is

$$r_n = \frac{nh}{2\pi m v_n} \qquad (2\text{-}23)$$

and substituting Equation (2-21) for v_n gives

$$r_n = \frac{h^2}{4\pi^2 e^2 m} n^2 \qquad (2\text{-}24)$$

Bohr used known values to calculate the size of the hydrogen atom in its stable ground state ($n = 1$). The result, 0.55×10^{-10} meter, was in agreement with observations of atomic size. Modern measurements yield a Bohr radius of 0.529×10^{-10} meter.

Calculating the hydrogen energy levels
Using the Bohr radius calculation and the associated velocity in the above equations, the total mechanical energy of the atom can be calculated. The potential energy, U, of the two electric charges separated by distance r_n is

$$U = \frac{-e^2}{r_n} \qquad (2\text{-}25)$$

and the negative sign ensures that the potential energy is highest at $r = \infty$. The kinetic energy is

$$K = \frac{1}{2}mv_n^2 \tag{2-26}$$

and substituting Equation (2-21) for the total energy

$$U + K = \frac{1}{2}m\left(\frac{e}{\sqrt{mr_n}}\right)^2 - \frac{e^2}{r_n} = -\frac{1}{2}\frac{e^2}{r_n} \tag{2-27}$$

so substituting the Bohr radius the energy for the nth level is shown to be

$$E_n = -\frac{2\pi^2 e^4 m}{h^2}\frac{1}{n^2} \tag{2-28}$$

Comparing Equation (2-28) to the experimentally determined Equation (2-17) allows the calculation of the Rydberg constant, R:

$$R = -\frac{2\pi^2 e^4 m}{h^3} \tag{2-29}$$

Bohr used the available experimental measurements for charge and mass of the electron (e, m) and Plank's constant h and calculated R = 3.1×10^{15} which was within the experimental errors of the best measurements of the Rydberg constant at the time.

Pickering Lines
Astronomers had observed atomic spectra in star light, called the Pickering lines, which had been believed to be hydrogen spectra; however, Bohr was able to show using that they were singly ionized helium spectra. A single electron atom, like helium, with nuclear charge $+Ze$ has energy levels for the single electron

$$E_n = -\frac{2\pi^2 e^4 Z^2 m}{h^2}\frac{1}{n^2} \tag{2-30}$$

which comes from the same derivation leading to Equation (2-28), but the Coulomb force is increased by Z ($Z = 2$ for helium). Using this equation a value of the Rydberg constant for singly ionized helium, R_{He} can be derived, and the ratio of $R_{He}/R = 4$.

de Broglie waves
It is possible in light of the wave-particle duality of matter to derive the hydrogen energy levels using the wave-like characteristics of the electron. The electron, supposed

to orbit the hydrogen nucleus at a radius, r_n, require that the circumference traveled by the electron must be an integer multiple of the electron wavelength, λ

$$2\pi r_n = n\lambda = n\frac{h}{mv_n} \tag{2-31}$$

and the wavelength is related to the electron's momentum by the de Broglie relation. This leads to the Bohr radius of Equation (2-24) and the Bohr energy levels. The reason that the circumferential distance must equal an integer value of the electron wavelength is otherwise the wave would have destructive interference. Another way of viewing it is that the wave must be continuous to be a stable wave, and continuity is achieved by Equation (2-31).

Niels Bohr published his model of the hydrogen atom in 1913 and was a triumph of the new quantum physics. However, it was not readily generalized to more complicated atoms and was replaced by quantum mechanics which is the subject of the next chapter.

References

Gamow, G., *Thirty Years That Shook Physics*, Dover Publications, Inc., 1966.

Halliday, D. and Resnick, R., *Fundamentals of Physics*, 2nd Ed., John Wiley & Sons, 1981.

Pais, A., Niels Bohr's Times, In Physics, Philosophy, and Polity, Clarendon Press, 1991.

Chapter 3 : Introduction to Quantum Mechanics

Introduction

In this chapter, the Schrödinger equation is described as the logical and more mathematical extension of de Broglie waves. An overview of wave equations is presented, and the meaning of vibrational normal modes is discussed because the Schrödinger equation is a wave equation. The solutions to the Schrödinger equation describe the orbitals for the atom, and the orbitals of the hydrogen atom are developed in detail.

Erwin Schrödinger

Austrian physicist Erwin Schrödinger quickly followed up on de Broglie's concept of a particle waves, and he developed a general equation for the description of de Broglie waves. He published what is now known as the *Schrödinger equation* in 1926. The Schrödinger equation is a differential equation very similar to the *wave equation* for other physical phenomena, such as vibrations on a string, acoustic waves, or electromagnetic waves. There are many analogies between a vibrating string and the Schrödinger equation, so the wave equation of the vibrating string is developed below.

Wave functions of a vibrating string

A vibrating string, such as a guitar string is described by the one dimensional wave equation. The string is under tension, T, and has mass density per unit length of ρ. The wave equation is describes the motion, or displacement of the string from its stationary position, and is a differential equation

$$\frac{\partial^2 \Psi}{\partial x^2} = \frac{\rho}{T} \frac{\partial^2 \Psi}{\partial t^2} = \frac{1}{c^2} \frac{\partial^2 \Psi}{\partial t^2} \qquad (3\text{-}1)$$

where Ψ is the transverse displacement of the guitar string in units of distance. The speed of propagation of the waves on the string is

$$c = \sqrt{\frac{T}{\rho}} \qquad (3\text{-}2)$$

Dimensional analysis can be used to show that the ratio of tension over string density is indeed wave speed squared. Using m.k.s units, tension over string density has units of force [newtons] over [kilograms/meter].

$$\left[\frac{N}{kg/m}\right] = \left[\frac{kg \cdot m/s^2}{kg/m}\right] = \left[\frac{m^2}{s^2}\right] \qquad (3\text{-}3)$$

Solutions to the wave equation for the vibrating string are *stationary vibrating modes* of the equation. The stationary modes must meet the *boundary conditions* of the string which do not allow displacement at the ends of the string where it is pinned. In the case of the guitar string, one end could include the fret, and the other end is pinned at the guitar bridge. If the length of the string is L, then the boundary conditions for the string displacement are

$$\Psi(0,t) = \Psi(L,t) = 0 \tag{3-4}$$

The boundary conditions are a mathematical description of the requirement that string cannot be displaced at the ends where it is pinned.

A family of solutions to the wave equation for the string vibration is

$$\Psi(x,t) = A \sin\left(2\pi \frac{x}{\lambda}\right) \sin(2\pi f t) \quad \lambda = 2L, L, \frac{2L}{3}, \frac{L}{2}, \dots \frac{2L}{N} \tag{3-5}$$

where A is the displacement amplitude, λ are the allowable wavelengths, and f is the frequency. Note that allowable wavelengths are $2L/N$ where N is an integer. Calculation of the spatial derivative of the solution for use in the wave equation (Equation (3-1)) gives

$$\frac{\partial^2 \Psi}{\partial x^2} = \left(\frac{2\pi}{\lambda}\right) \frac{\partial}{\partial x}\left[A \cos\left(2\pi \frac{x}{\lambda}\right)\right] \sin(2\pi f t) \tag{3-6}$$

$$= -\left(\frac{2\pi}{\lambda}\right)^2 \left[A \sin\left(2\pi \frac{x}{\lambda}\right)\right] \sin(2\pi f t)$$

$$= -\left(\frac{2\pi}{\lambda}\right)^2 \Psi(x,t)$$

Thus the spatial and time derivatives are shown to be

$$\frac{\partial^2 \Psi}{\partial x^2} = -\left(\frac{2\pi}{\lambda}\right)^2 \Psi(x,t) \tag{3-7}$$

$$\frac{\partial^2 \Psi}{\partial t^2} = -(2\pi f)^2 \Psi(x,t) = -\left(2\pi \frac{c}{\lambda}\right)^2 \Psi(x,t)$$

and thus the wave equation is shown to hold for the solutions in Equation (3-5).

Stationary modes and wave frequency

The most general solution to the wave equation includes an arbitrary phase, ϕ, which is added to the time dependent part of the solution for Ψ. The solutions to the wave equation are partitioned into a stationary mode, which is only spatially dependent (Equation (3-8)).

$$\Psi(x,t) = A \sin\left(2\pi \frac{x}{\lambda}\right) \sin(2\pi f t + \phi) \quad \lambda = 2L, L, \frac{2L}{3}, \frac{L}{2}, \ldots \frac{2L}{N} \qquad (3\text{-}8)$$

This stationary mode is multiplied by a periodic time varying function. The spatial distribution of the stationary modes is what determines the important aspects of the guitar string vibration, like the frequency (see Figure 2-4). Likewise, as we'll show below the stationary modes of the wave functions in the atom determine its important properties.

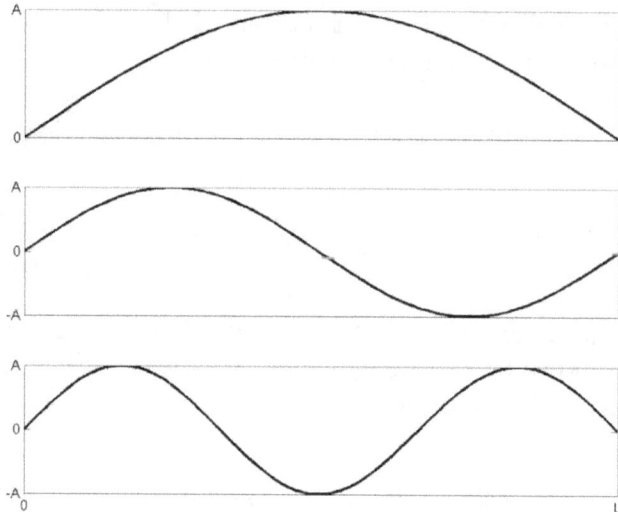

Figure 3-1: Stationary vibrational modes for one-dimensional string. The normal modes must meet the boundary conditions of the wave function, so they cannot have any displacement at the ends (0 or L) where the string is pinned. The figure illustrates the first three vibrational stationary modes of the string.

The frequency of the vibrating string, using Equation (3-8) is

$$f = \frac{c}{\lambda} = \sqrt{\frac{T}{\rho}} \frac{1}{\lambda} \quad \lambda = 2L, L, \frac{2L}{3}, \frac{L}{2}, \ldots \frac{2L}{N} \qquad (3\text{-}9)$$

By comparing this calculation with real world observation, we can see that the result for the frequency in Equation (3-9) provides some intuitive conclusions. If the vibrating string is a guitar string, then vibrational frequency of the fundamental stationary mode ($\lambda = 2L$) is the pitch of the string when strummed. The pitch increases with the square root of the string tension, which is consistent with the observation that tightening the string increases the pitch. The pitch is inversely proportional with the string length (L) which is consistent with the observation that using a higher fret, which shortens the string, increases the pitch.

Schrödinger equation

The Schrödinger equation is a wave equation like a wave on string except it is a description of the wave which characterizes the electron in an atom. For atoms, the Schrödinger wave function will fully characterized the atomic energy states. In this chapter we will discuss the single electron atom, like a hydrogen atom or singly ionized helium atom. Since the atom is a three dimensional object, the wave function is also three dimensional. One can visualize an analogy with acoustic vibrations inside a hard metal shell. In the case of an acoustic wave the sound wave is vibrating the air inside the shell. What exactly is vibrating in the case of the wave function for an atom will be shown later.

Solutions to the Schrödinger equation for one-electron atoms

The Schrödinger equation solutions are the wave functions for one-electron atoms. These solutions produce the same energy levels as the Bohr energy levels. The electron binding energies, E_n, are

$$E_n = -\frac{Z^2 hR}{n^2} \tag{3-10}$$

which are determined by the *principal quantum number n*, and Plank's constant, h, the Rydberg constant, R and the atomic number Z.

The three dimensional Schrödinger equation requires some new notation. The sum of the second derivatives of the *Schrödinger wave function* is shown in compact form using the *Laplace operator*. The Laplace operator is a second order differential operator in three dimensional space

$$\nabla^2 \Psi = \frac{\partial^2 \Psi}{\partial x^2} + \frac{\partial^2 \Psi}{\partial y^2} + \frac{\partial^2 \Psi}{\partial z^2} \tag{3-11}$$

where the left hand side is pronounce "del-squared". The second derivatives of the wave function, Ψ, are over the three spatial dimensions, x, y, and z. The Schrödinger equation, for the stationary vibrational modes is

$$-\frac{h^2}{8\pi m}\nabla^2 \Psi + V(r)\Psi = E_n \Psi \tag{3-12}$$

Each term in Equation (3-12) requires explanation.

- It is assumed that the spatial derivatives (∇^2) are over three Cartesian spatial dimensions x, y, z or over spherical coordinates r, θ, and ϕ. For the atom spherical coordinates are more convenient because many solutions have spherical symmetry (described below).
- The binding energies, E_n, are the Bohr binding energies.
- The *potential energy* of the atom, $V(r)$, is the Coulomb potential energy.
- h is Plank's constant, m is electron mass.

Equation (3-12) is the *time independent Schrödinger equation*, where the variable Ψ is the Schrödinger wave function. The wave function fully characterizes the atomic properties. As described above, the wave function is more easily understood when represented in the spherical coordinates. The wave functions for one electron atoms partition into separable functions:

$$\Psi(x,y,z) = \Psi(r,\theta,\phi) = R(r)\Theta(\theta)\Phi(\phi) \qquad (3\text{-}13)$$

s-orbitals: spherically symmetric wave functions

The time independent solutions are the stationary modes analogous to the stationary vibrational modes of the waves on a vibrating guitar string. Each energy level, E_n, has wave functions that are spherically symmetric, meaning that it does not depend upon θ, and ϕ. The wave function depends only upon the radial variable, r. The first example of a spherically symmetric wave function is the ground state. The ground state is referred to as Ψ_{100}:

$$\Psi_{100} = \frac{1}{\sqrt{\pi}} \left(\frac{Z}{a_0}\right)^{3/2} e^{-Zr/a_0} \qquad (3\text{-}14)$$

The subscript numbers 100 refer to the principal quantum number (n=1) and the two other quantum numbers ($\ell=0$, $m=0$). The other quantum number, ℓ and m, will be discussed in more detail below. A plot of the ground state wave function is shown in Figure 3-2 (left side) which illustrates the behavior of the exponential function in Equation (3-14). The exponential function rapidly decreases in value toward zero as r increases.

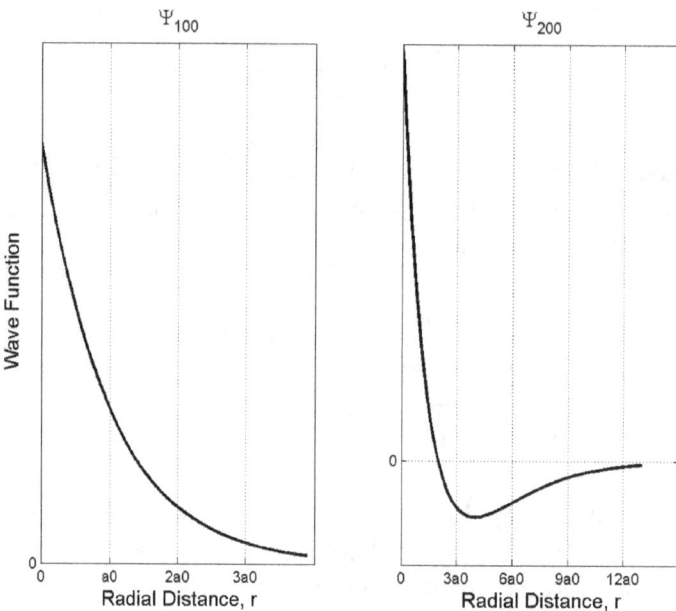

Figure 3-2: Spherically symmetric wave functions which correspond to s-orbitals. The ground state wave function is shown on the left side, and the first excited state is on the right side. The term a0 represents the Bohr radius, which is a constant. Note that the value of the wave function can be positive or negative.

The term a_0 is the Bohr radius for the ground state. The *Bohr radius for hydrogen* in m.k.s units is

$$a_0 = \frac{\varepsilon_0 h^2}{4\pi^2 e^2 m} = 0.529 \times 10^{-10} \text{ m} \qquad (3\text{-}15)$$

The Ψ_{100} wave function is referred to as the 1s orbital. The s orbitals are the spherically symmetric wave functions. The first excited state (n=2) has a 2s orbital (Ψ_{200}):

$$\Psi_{200} = \frac{1}{4\sqrt{2\pi}} \left(\frac{Z}{a_0}\right)^{3/2} \left(2 - \frac{Zr}{a_0}\right) e^{-Zr/2a_0} \qquad (3\text{-}16)$$

The subscript numbers 200 refer to the principal quantum number (n=2) and the two other quantum numbers (ℓ=0, m = 0). The Ψ_{200} wave function is plotted in Figure 3-2. The constant a_0 has the same value as the Bohr radius for the ground state. Note that Ψ_{200} has values that are positive or negative. The position where the wave function crosses zero is called a *node*. The wave function is generally complex valued, and can take on signed values like other kinds of wave functions (see waves on a string). The wave functions for higher excited s orbitals are shown in the table at the end of this chapter.

Interpretation of the wave function

Since the Schrödinger wave function is a solution to a wave equation, the question naturally arises about what is vibrating. The answer is that nothing is actually vibrating. This is where the analogy with other wave functions breaks down. For an electromagnetic wave, the wave function that describes it predicts the number of photons observed. The Schrödinger wave function does not determine particles or measurements in the same way.

It is worth quoting the physicist George Gamow at length, for the colorful description if nothing else. He explains the interpretation of wave functions, or ψ-functions as he refers to them:

"Since in atomic and nuclear physics the notion of classical linear trajectories inevitably fails, it is apparently necessary to devise another method for describing the motion of the material particles, and here the ψ-functions come to our aid. They do not represent any physical reality. The de Broglie waves have no mass such as we find in the case of electromagnetic waves, and whereas, in principle, one can buy half a pound of red light, there does not exist in the world an ounce of de Broglie waves. They are no more material than the linear trajectories of classical mechanics, and, in fact, can be described as "widened mathematical lines." They guide the motion of particles in quantum mechanics in the same sense as the linear trajectories guide the motion of particles in classical mechanics. But, just as we do not consider the orbits of planets in the Solar System as some kind of railroad tracks that force Venus and Mars and our own Earth to move along elliptical orbits, we may not consider the wave-mechanical continuous functions as some field of forces which influences the motion of electrons. The de Broglie- Schrödinger wave functions (or, rather, the square of their absolute values, i.e. $|\psi|^2$) just determine the probability that the particle will be found in one or another part of space and will move with one or another velocity.

The German physicist Max Born formulated the modern interpretation of the wave function, and the interpretation for the single electron atom will be determined in detail.

Born's formulation is the *probability density function* description of the wave function. Mathematically, the probability density is determined to be the square of the magnitude of the wave function which is written as $\Psi^*\Psi$.

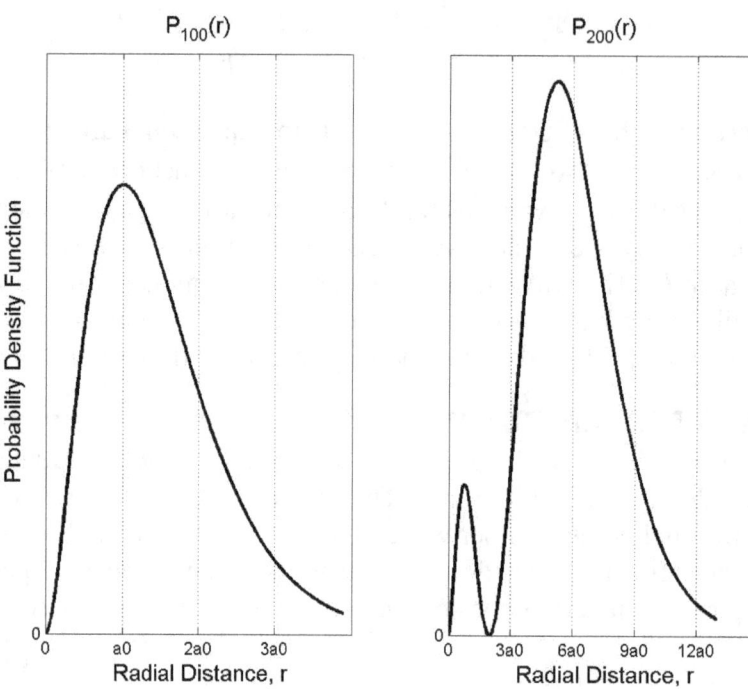

Figure 3-3: Radial probability density functions (PDF) for the first two s-orbitals. The ground state PDF is shown on the left side, and the first excited state is on the right side. Note that the peak value for the ground state coincides with the Bohr radius, a0, which is a constant.

Probability density functions

In the case of the *s* orbitals of single electron atom the probability distribution for the electron at an orbital radius *r*, can be characterized by the *probability density function*, or PDF, *P(r)* which is defined as the probability density of finding the electron between radius *r* and *r* + *dr*, where *dr* is the infinitesimal derivative of *r*. This implies some properties of the *radial probability density function*, *P(r)*:

$$P(r)dr = \text{probability of finding electron between } r \text{ and } r + dr \qquad (3\text{-}17)$$

$$P(r) \geq 0$$

$$\int_0^\infty P(r)dr = 1$$

The first statement is a consequence of the definition. Since the PDF is a probability density in dimension *r*, multiplying by the infinitesimal distance *dr* gives a probability. The infinite integral follows from the meaning of probability, since the probability of finding the electron over all space is 1. As stated above the probability density for the electron in the atom, which is probability per unit volume, is given by the square of the magnitude of the wave function. The radial PDF is calculated as follows:

$$|\Psi(r)|^2 \quad \times \quad 4\pi r^2 dr = P(r) \qquad (3\text{-}18)$$

$$\frac{\text{probability}}{\text{volume}} \times \text{volume} = \text{radial PDF}$$

The importance of the normalization factors in the wave functions in Equations (3-14) and (5-2) is now clear. The normalization factors are chosen so that

$$\int_0^\infty |\Psi(r)|^2 4\pi r^2 dr = 1 \qquad (3\text{-}19)$$

The PDF for the electron is for 3-D space, thus the integrand is the probability of finding the electron in a spherical shell of thickness dr at radius r. For spherically symmetric wave functions only (s orbitals), multiplying the wave function squared by the term $4\pi r^2$ produces the radial PDF. The radial PDF functions, $P(r)$, for the first two s orbitals, Ψ_{100} and Ψ_{200} are shown in Figure 3-3. In the plots, note that the PDF value is always non-negative even in cases where the wave function was negative. For the ground state, the peak value of the radial PDF is a_0, the Bohr radius. For the first excited state, Ψ_{200}, the peak value of the radial PDF is nearly $6a_0$, so the first excited state is quantitatively larger than the ground state.

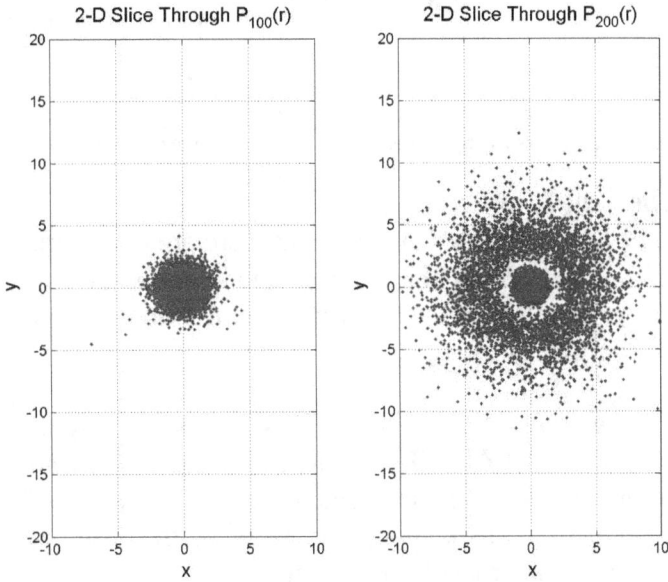

Figure 3-4: Slices through the electron probability density functions (PDF) represented as an electron cloud. Each point represents 1/10,000th of the probability of finding an electron at that position. The 1s orbital is on the left and the 2s orbital on the right. The units in the x and y dimension are in terms of the Bohr radius a_0 (1 is one Bohr radius).

One consequence of quantum mechanics is that the electron in a hydrogen atom is no longer visualized as having well defined circular orbits as in Bohr's model. The PDF can be visualized as describing an electron cloud in which the electron has a more probability where the cloud is most dense. A representation of the electron cloud for the 1s and 2s orbitals (P_{100} and P_{200}) is shown in Figure 3-4. The PDF functions are represented as an electron cloud. A 2-dimensional slice through the 3-dimensional PDF is shown in which each point represents 1/10,000th of the probability of finding an electron at that position. It can be seen that the atom is physically larger in radius for the first excited state, which is expected. However, the dimension or size of the hydrogen atom is a probabilistic concept rather than a deterministic concept as size would be for a macroscopic object like a ball.

	State label	Wave function	Orbital	E_n
n=1 ℓ=0 m=0	100	ψ_{100}	1s	$-hR/1^2$
n=2 ℓ=0 m=0	200	ψ_{200}	2s	$-hR/2^2$
n=2 ℓ=1 m=+1	211	ψ_{211}	$2p_x$ (or $2p_y$)	$-hR/2^2$
n=2 ℓ=1 m=0	210	ψ_{210}	$2p_z$	$-hR/2^2$
n=2 ℓ=1 m=-1	21-1	ψ_{21-1}	$2p_y$ (or $2p_x$)	$-hR/2^2$

Table 3-1: Quantum numbers for hydrogen orbitals and the energy associated with the orbital.

Quantum numbers n, ℓ, m

It was mentioned earlier in this chapter that there are quantum number other than the principal quantum number which characterize and describe electron orbitals. The other quantum numbers are represented by the variables n, ℓ and m. As stated above, n is the *principal quantum number*.

The *angular momentum quantum number* is represented by ℓ and has values which are related to n. The values of ℓ are 0,1,2...n-1. The value of ℓ implies increasing kinetic energy, and therefore greater angular momentum. If the value of ℓ was equal to n, it would imply that all energy was kinetic, and therefore the atom has no potential energy, so ℓ cannot be equal to n, and must be less than n.

The *magnetic quantum number* is represented by m and has values related to ℓ. The values of m are -ℓ, -ℓ+1,..., 0, ..., ℓ-1, ℓ. The value of m represents the z-component of angular momentum, where z is the vertical Cartesian spatial dimension.

Generally, the wave function, Ψ, is a function of the spherical coordinates r, θ, and ϕ, and is parameterized by the three quantum numbers n, ℓ, m. The wave function matches the orbital designation.

Shell and subshell description

Often the electron wave functions are described using the shell/subshell description. The *shell* the electron occupies is determined by the principal quantum number n. The *subshell* comes about in the multi-electron atom because the electron-electron repulsion causes the subshells to have different energy levels. Within the shell there can be n subshells. The subshells are distinguished by their angular momentum quantum number, ℓ. In a single electron atom all the orbitals which have the same principal quantum number n have the same binding energy. Orbitals with equal binding energy are referred to as *degenerate orbitals*.

Examples of the shells and subshells for the hydrogen atom are shown in Table 3-2. Shells are described here using n, the subshell is ℓ but is typically described using the letters *s*, *p*, *d*, and *f*. Table 3-2 gives descriptions of the orbitals for the first 4 shells. For the first shell, ℓ can only have value of 0, so only one subshell is possible. For $n=3$, there are three values of ℓ (0, 1, 2) which implies shell 3 can have 3 subshells *s*, *p* and *d*.

Shell	Subshell	Orbital
$n=4$	$\ell=3, f$	$m=3,2,1,0,-1,-2,-3$
	$\ell=2, d$	$(4f)$
	$\ell=1, p$	$m=+2,+1,0,-1,-2$
	$\ell=0, s$	$(4d)$
		$m=+1,0,-1$ $(4p)$
		$m=0$ $(4s)$
$n=3$	$\ell=2, d$	$m=+2,+1,0,-1,-2$
	$\ell=1, p$	$(3d)$
	$\ell=0, s$	$m=+1,0,-1$ $(3p)$
		$m=0$ $(3s)$
$n=2$	$\ell=1, p$	$m=+1,0,-1$ $(2p)$
	$\ell=0, s$	$m=0$ $(2s)$
$n=1$	$\ell=0, s$	$m=0$ $(1s)$

Table 3-2: Shell and subshell designation for hydrogen (or single electron atoms).

p orbitals for the single electron atom

The orbital notation matches the subshell notation (s, p, d, f) where

- $\ell=0$ indicates an *s* orbital
- $\ell=1$ *p* orbital
- $\ell=2$ *d* orbital
- $\ell=3$ *f* orbital

The magnetic quantum number, m determines orientation of the wave functions in Cartesian coordinates *x*, *y*, and *z* which will be discussed below. The convention is to describe the orbital with the appropriate letter, such as *p*, and the subscript *x*, *y* or *z* depending upon m. For example:
- $\ell=1$, m=0 indicates a p_z orbital

- $\ell=1$, $m=\pm 1$ indicates p_x or p_y orbital

Depending upon convention, the p_x and p_y orbitals can be interchanged; however, $m=0$ is always referred to as p_z. Examples of orbital notation for hydrogen are shown in Table 5-1. All the orbitals which have the principal quantum number $n=2$ are degenerate, *i.e.*, they have the same binding energy.

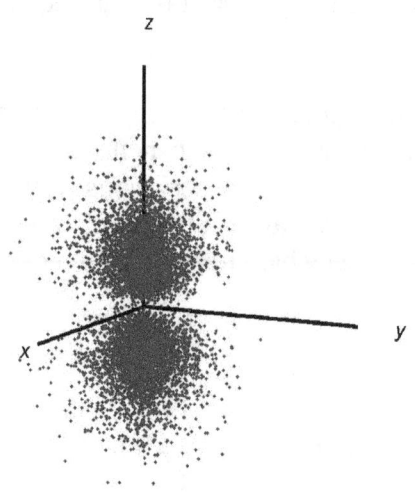

Figure 3-5: Electron cloud representation of $2p_z$ orbital for hydrogen. Note the *x-y* plane is the nodal plane.

p Orbital Wave Functions for Single Electron Atoms

For any subshell, $\ell=1$, there are 3 p orbitals corresponding to $m = 0, 1, -1$ and the shape of the p orbitals have angular dependence unlike the spherically symmetric s orbitals. For example the p_z orbital is mathematically described by the wave function Ψ_{210}:

$$\Psi_{210} = \frac{1}{4\sqrt{2\pi}} \left(\frac{Z}{a_0}\right)^{3/2} \frac{Zr}{a_0} e^{-Zr/2a_0} \cos\theta \qquad (3\text{-}20)$$

Note that it does depend upon the radial variable, r and the angle θ. The subscript number 210 refer to the principal quantum number ($n=2$) and the two other quantum numbers ($\ell=1$, $m=0$).

An electron cloud representation of the p_z orbital is shown in Figure 3-5. Note that the shape of the orbital clearly does not have spherical symmetry; it has angular and radial dependence. There is a *nodal plane* for each of the 2p orbitals. A nodal plane is plane where the wave function, and therefore the probability density, is equal to zero. For the $2p_z$ orbital, the *x-y* plane is the nodal plane.

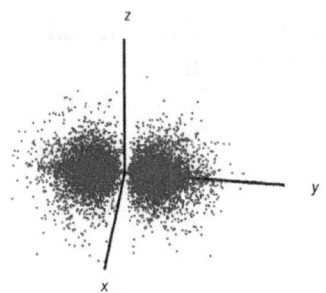

Figure 3-6: Electron cloud representation of $2p_y$ orbital for hydrogen. Note the x-z plane is the nodal plane.

The p_x and p_y orbitals are mathematically described by the following wave function $\Psi_{21\pm1}$:

$$\Psi_{21\pm1} = \frac{1}{8\sqrt{\pi}}\left(\frac{Z}{a_0}\right)^{3/2}\frac{Zr}{a_0}e^{-Zr/2a_0}\sin\theta \cdot e^{\pm i\phi} \tag{3-21}$$

Note that these wave functions depend upon the radial variable, r and both angles θ and ϕ. The subscript numbers 21±1 refer to the principal quantum number ($n=2$) and the two other quantum numbers ($\ell=1$, $m=\pm1$). An electron cloud representation of the p_y (or p_x) orbital is shown in Figure 3-6. The axes have been rotated to more clearly observe that the x-z plane is the nodal plane for this orbital. The three $2p$ orbitals are perpendicular to each other. Each $2p$ orbital is aligned with one of the x, y or z axes.

Radial Probability Density Functions for p orbitals

Using the method in Equation (3-18), the radial probability density functions, PDFs, for the p orbitals can be calculated. The radial PDF of the $2s$ and $2p$ orbitals for the hydrogen atom are plotted in Figure 3-7. By plotting the radial PDF functions, the size of the atom is apparent. It is very clear that the $2s$ and $2p$ orbitals for the hydrogen atom are much larger than the $1s$ orbital. It can also be observed that the peak value of PDF for the $2p$ orbital is at shorter range, r, than the peak of the $2s$ orbital. In fact, p orbitals are smaller than s orbitals. Another way to visualize this is to say the electron is closer to the center of the atom for p orbitals than for s orbitals. Generally, as ℓ increases for given n, the atomic size decreases.

Also, note that there is one node, where the wave function is zero for the $2s$ orbital, and there are not any nodes for the $2p$ orbital. The zero value for the PDF at $r = 0$ is not due to a node in the wave function; it is due to the multiplicative term $4\pi r^2$ in Equation (3-18). So, the zero value at $r=0$ is not a node.

Generally, a wave function will have *n*-ℓ-1 radial nodes. Can we predict the number of nodes for the 3*s* and 3*p* orbitals? Using the quantum number for this example:

- 3*s* orbital: *n*=3, ℓ=0, there will be 2 radial nodes
- 3*p* orbital: *n*=3, ℓ=1, 1 radial node

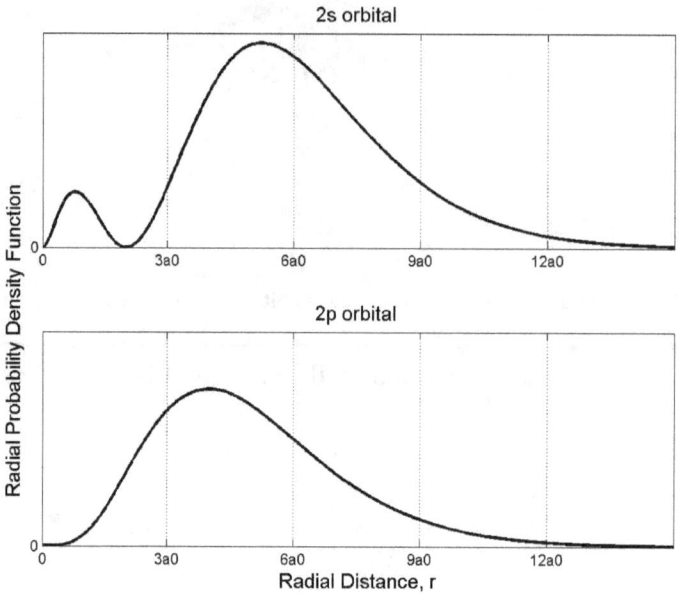

Figure 3-7: Radial probability density functions for the 2*s* and 2*p* orbitals of the hydrogen atom. The term a0 represents the Bohr radius, which is a constant. Note the node for the 2*s* orbital at approximately a0 where the wave function is zero.

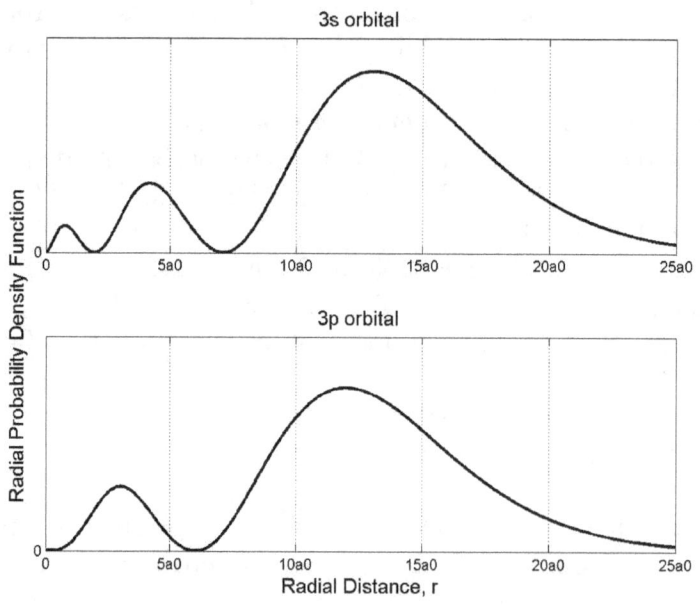

Figure 3-8: Radial probability density functions for the 3*s* and 3*p* orbitals of the hydrogen atom. The 3*s* orbital has 2 nodes and the 3*p* orbital has one node.

The radial PDF of the 3s and 3p orbitals for the hydrogen atom are plotted in Figure 3-8. It is easily observed that the 3s orbital has 2 radial nodes, and the 3p orbital has 1 radial node. Consistent with the statement that atomic size decreases with increasing value of ℓ, the 3p orbital is smaller than the 3s orbital. This is because the peak value of PDF for the 3s orbital is at greater range, r, than the peak of the 3p orbital.

The expected radius for an electron, often referred to as the *mean radius* since it is the average radius calculated using the radial PDF, can be calculated:

$$\bar{r} = \frac{n^2 a_0}{Z}\left[1 + \frac{1}{2}\left(1 - \frac{\ell(\ell-1)}{n^2}\right)\right] \tag{3-22}$$

Note that Bohr's calculation for the quantized radius for the hydrogen atom, derived in Chapter 2 is

$$r_n = \frac{a_0}{Z} n^2 \tag{3-23}$$

which is not consistent with the Schrödinger equation solutions. So quantum mechanics provides an updated model which is more accurate than the Bohr model.

Quantum numbers			Wave Function
N	ℓ	M	
1	0	0	$\Psi_{100} = \dfrac{1}{\sqrt{\pi}}\left(\dfrac{Z}{a_0}\right)^{3/2} e^{-Zr/a_0}$
2	0	0	$\Psi_{200} = \dfrac{1}{4\sqrt{2\pi}}\left(\dfrac{Z}{a_0}\right)^{3/2}\left(2 - \dfrac{Zr}{a_0}\right) e^{-Zr/2a_0}$
2	1	0	$\Psi_{210} = \dfrac{1}{4\sqrt{2\pi}}\left(\dfrac{Z}{a_0}\right)^{3/2} \dfrac{Zr}{a_0} e^{-Zr/2a_0} \cos\theta$
2	1	±1	$\Psi_{21\pm 1} = \dfrac{1}{8\sqrt{\pi}}\left(\dfrac{Z}{a_0}\right)^{3/2} \dfrac{Zr}{a_0} e^{-Zr/2a_0} \sin\theta \cdot e^{\pm i\phi}$
3	0	0	$\Psi_{300} = \dfrac{1}{81\sqrt{3\pi}}\left(\dfrac{Z}{a_0}\right)^{3/2}\left(27 - 18\dfrac{Zr}{a_0} + 2\dfrac{Z^2 r^2}{a_0^2}\right) e^{-Zr/3a_0}$
3	1	0	$\Psi_{310} = \dfrac{1}{81}\sqrt{\dfrac{2}{\pi}}\left(\dfrac{Z}{a_0}\right)^{3/2}\left(6 - \dfrac{Zr}{a_0}\right)\dfrac{Zr}{a_0} e^{-Zr/3a_0} \cos\theta$
3	1	±1	$\Psi_{31\pm 1} = \dfrac{1}{81\sqrt{\pi}}\left(\dfrac{Z}{a_0}\right)^{3/2}\left(6 - \dfrac{Zr}{a_0}\right)\dfrac{Zr}{a_0} e^{-Zr/3a_0} \sin\theta \cdot e^{\pm i\phi}$
3	2	0	$\Psi_{320} = \dfrac{1}{81\sqrt{6\pi}}\left(\dfrac{Z}{a_0}\right)^{3/2} \dfrac{Z^2 r^2}{a_0^2} e^{-Zr/3a_0}(3\cos^2\theta - 1)$
3	2	±1	$\Psi_{32\pm 1} = \dfrac{1}{81\sqrt{\pi}}\left(\dfrac{Z}{a_0}\right)^{3/2} \dfrac{Z^2 r^2}{a_0^2} e^{-Zr/3a_0} \sin\theta\cos\theta \cdot e^{\pm i\phi}$
3	2	±2	$\Psi_{32\pm 2} = \dfrac{1}{162\sqrt{\pi}}\left(\dfrac{Z}{a_0}\right)^{3/2} \dfrac{Z^2 r^2}{a_0^2} e^{-Zr/3a_0} \sin^2\theta \cdot e^{\pm i 2\phi}$

Table 3-3: Wave functions for single electron atoms. The parameter a_0 is the Bohr radius, Z is the atomic number. The wave functions are represented in terms of the spherical coordinates: r, θ, ϕ.

References

Gamow, G., *Thirty Years That Shook Physics*, Dover Publications, Inc., 1966.

Eisner, R. and Resnick, R., Quantum Physics of Atoms, Molecules, Solids, Nuclei, and Particles, John Wiley & Sons, 1985.

Sienko, M.J. and Plane, R.A., *Chemistry, Principles and Applications*, McGraw-Hill Book Company, 1979.

Chapter 4 : Multi-Electron Atoms

Introduction

In this chapter, we use the principles that we developed for the single electron atom and apply these to multi-electron atoms. We note the differences that arise in a multi-electron atom to electron interaction. In particular, we introduce the concept of electron spin and how demonstrate how important that concept is in the electron and orbital structure of multi-electron atoms. We describe the periodic properties of atoms: atomic radius, ionization energy and electron affinity, and we show the connection between quantum mechanics and the periodic physical properties.

Electron spin

The quantum mechanical description for a single electron atom parameterizes each atomic *orbital* by the three quantum numbers n, ℓ, and m. With multiple electron atoms, each of the electrons in the atom is said to occupy a *state* which is characterized by *four* quantum numbers. The fourth quantum number is called the *spin magnetic quantum number*, m_s. This fourth quantum number takes on two values: $\pm\frac{1}{2}$. The first four quantum numbers are properties of the orbital which describe the electron's position in three dimensions. The fourth quantum number, m_s, is a property of the electron, not the orbital.

Meaning of spin

It's important to understand that there is no classical physics analogy to spin. In the discussion that follows, the spin will be visualize in terms of classical physics to some extent, but it is not to be taken literally. When the quantum number m_s has values of $+\frac{1}{2}$, the electron is said to be in a "spin up" state, and when $m_s = -\frac{1}{2}$, it is said to be "spin down." The electron spin is the projection of the electron's angular momentum in the z-direction. The electron is often visualized as a sphere spinning on its axis which provides angular momentum. Thus, the terms spin up and spin down can be visualized as the electron rotating in opposite directions. The direction of the spin angular momentum can change, but the magnitude of the angular momentum is unchanging. For this reason, the electron spin is an intrinsic property, like the electron's mass or charge. Spin is often referred to as intrinsic spin in order to make that clear.

The electron is often visualized as spinning mass, but this is a conceptual picture only. The most recent experimental evidence indicates the electron is a *point particle*, meaning it has no physical extent in space. If the electron had appreciable extent in space, then the scattering forces between multiple electrons would need to be modified, in a similar manner to the scattering of an alpha particle by a nucleus in Rutherford's experiment. However, in experiments which studied high energy electron-electron scattering, even the closest collisions indicate that electrons are properly viewed as point particles. So we refer to the electron's intrinsic spin, but we cannot claim that it is due to the electron actually spinning.

Goudsmit and Uhlenbeck

In 1925, two Dutch graduate students, Samuel Goudsmit and George Uhlenbeck, first elucidated the physical nature of the fourth quantum number. Three years earlier, the Stern-Gerlach experiment (described below) demonstrated that atoms had quantized angular momentum, but the significance of the experimental results for intrinsic spin had not been realized. Goudsmit and Uhlenbeck were trying to understand why hydrogen and sodium optical spectra were composed of very closely spaced multiple lines. The physicist Wolfgang Pauli proposed that a fourth quantum number might be needed to explain these and other atomic spectral features, but he did not have a physical interpretation of the fourth quantum number. Uhlenbeck describes their thinking:

> *"Goudsmit and myself hit upon this idea by studying a paper of Pauli, in which the famous exclusion principle was formulated and in which, for the first time, four quantum numbers were ascribed to the electron. This was done rather formally; no concrete picture was connected with it. To us this was a mystery. We were so conversant with the proposition every quantum number corresponds to a degree of freedom, and on the other hand with the idea of a point electron, which obviously had three degrees of freedom only, that we could not place the fourth quantum number. We could understand it only if the electron was assumed to be a small sphere that could rotate."*

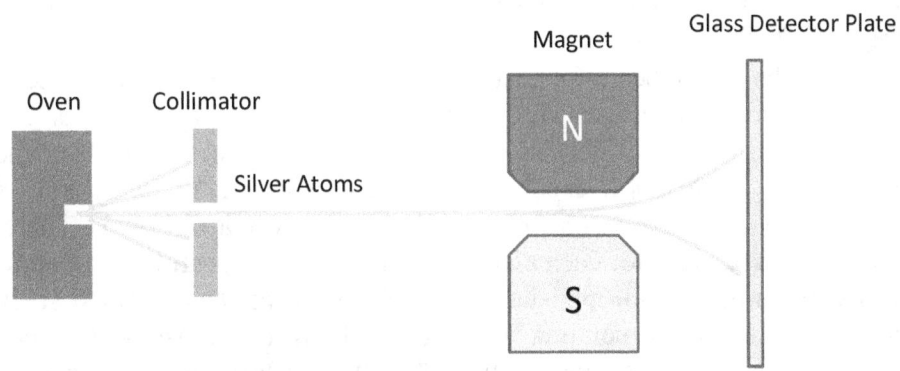

Figure 4-1: Diagram of the Stern-Gerlach experiment. A beam of neutrally charged silver atoms is formed by evaporating silver from an oven. A collimator, which is a small slit, narrows the beam. A magnet which provides a strong magnetic field gradient in the z-direction separates the beam into two beams which are detected by condensing on a glass plate.

Stern-Gerlach Experiment

An elegant experiment conducted by Otto Stern and Walther Gerlach in 1922 demonstrated that atoms had quantized angular momentum, which is due to the electron's intrinsic spin. The experiment, now known as the Stern-Gerlach experiment was conducted with silver atoms which provided the first experimental evidence of intrinsic spin. The intrinsic spin of the electron provides an angular momentum, and the spin quantum number, m_s, indicates the electron spin angular momentum in the z-direction.

The features of the Stern-Gerlach experiment are shown in Figure 5-1. An oven evaporates silver atoms, which leave the oven as electrically neutral atoms through narrow aperture in the oven. A beam of charged atoms will be displaced by the magnetic field, so a beam of electrically neutral atoms will eliminate that effect. A narrow slit in a barrier is a collimator which narrows the beam of silver atoms which will condense on a glass plate as a narrow spot. A powerful magnet provides a magnetic field *gradient* in the vertical, or z-direction, which splits the narrow beam into exactly two beams. The reason the beam is distorted by the magnetic field gradient in z is because the angular momentum of the electron, which is charged, creates a magnetic field emanating from the electron. The electron thus becomes a small magnetic dipole which depending upon its orientation is displaced in a magnetic field gradient.

N	ℓ	m	m_s
2	1	−1	−½
2	1	−1	+½
2	1	0	−½
2	1	0	+½
2	1	+1	−½
2	1	+1	+½

Table 4-1: Quantum numbers for the 2p electrons in the silver atom. Note there is no net angular momentum because for each electron with a non-zero angular momentum there is an electron with an equal and opposite angular momentum.

The reason the silver beam is separated into exactly two beams, is because of the electron distribution in the atom. The neutral silver atom has 47 electrons, and the first 46 electrons have orbital distributions such that there is no net angular momentum for these 46 electrons. An example is shown in Table 4-1 for the full $2p$ orbital. The quantum numbers for each of the $2p$ electrons in the silver atom are shown. There is no net angular momentum because for each electron with a non-zero angular momentum there is an electron with an equal and opposite angular momentum. Thus a filled shell and sub-shell have no net angular momentum. In the case of silver, the last electron occupies the $5s$ orbital, which has no angular momentum ($\ell = 0$); therefore, the angular momentum of the $5s$ electrons in the atom is determined entirely by its spin state. If the spin magnetic quantum number, m_s is +½, then the silver atom is displace by a force proportional to the electron intrinsic angular momentum. If $m_s = -½$, the electron is displaced by an equal and opposite force. Since the distribution of spin up and spin down electrons is random and equal, the beam of silver atoms is divided into two beams by the magnetic field gradient in z.

Pauli exclusion principle

Each electron in a multi-electron atom occupies a state characterized by four quantum numbers: n, ℓ, m and m_s. The physicist Wolfgang Pauli first discovered the exclusion principle in 1925 which now bears his name. The *Pauli exclusion principle* may be summarized as follows: *in a multi-electron atom each of the electrons in the atom must occupy a unique state, characterized by the four quantum numbers.* No two electrons in an atom can have the same four quantum numbers. The first three quantum numbers describe an orbital, and m_s describes the electron in an orbital. It follows that each orbital can have no more than two electrons, one with spin up ($m_s = +\frac{1}{2}$) and one with spin down ($m_s = -\frac{1}{2}$). A diagram for the neon and argon atoms is shown in

Figure 4-2. The orbital notation ($1s$, $2p_x$, etc.) is equivalent to writing the first three quantum numbers, and the arrows represent the spin up and spin down states for the electron. Neon has each of the orbitals filled with two electrons. Note that two electrons occupying the same orbital are required to have the opposite spin. Similarly, argon has each of its p orbitals filled.

$$\text{Ne} \quad \underset{1s^2}{\downarrow\uparrow} \quad \underset{2s^2}{\downarrow\uparrow} \quad \underset{2p_x^2}{\downarrow\uparrow} \quad \underset{2p_y^2}{\downarrow\uparrow} \quad \underset{2p_z^2}{\downarrow\uparrow}$$

$$\text{Ar} \quad \underset{1s^2}{\downarrow\uparrow} \quad \underset{2s^2}{\downarrow\uparrow} \quad \underset{2p_x^2}{\downarrow\uparrow} \quad \underset{2p_y^2}{\downarrow\uparrow} \quad \underset{2p_z^2}{\downarrow\uparrow} \quad \underset{3s^2}{\downarrow\uparrow} \quad \underset{3p_z^2}{\downarrow\uparrow} \quad \underset{3p_z^2}{\downarrow\uparrow} \quad \underset{3p_z^2}{\downarrow\uparrow}$$

Figure 4-2: Diagram representing the electron states for the neon and argon atoms. The electrons do not simply fill up the ground state, which is the lowest energy state due to the *Pauli exclusion principle*.

The Pauli exclusion principle is used to describe the order in which multi-electron atomic orbitals are filled. The exclusion principle prevents all of the electrons from occupying the lowest energy state, or ground state, which is the $1s$ orbital. Pauli realized that the exclusion principle was a property of electrons and not specifically atoms, so the exclusion principle is applicable to many different systems containing electrons. In addition, the exclusion principle has much broader application. It can be applied to all particles that have intrinsic spin characterized by two states, which are often called spin ½ particles. Protons, neutrons, and electrons are all spin ½ particles, so the exclusion principle also describes how states in the nucleus are occupied. This chapter will focus only on the application of the exclusion principle to electron states.

Wave functions for multi-electron atoms

The wave functions for the hydrogen atom, which are solutions to the Schrödinger equation, describe the single electron in three dimensions. When there are multiple electrons in an atom, electron-electron repulsion needs to be taken into account. Any multi-electron atom is accurately described using the Schrödinger equation; however, the Schrödinger equation and the solutions to it become more complicated with greater numbers of electrons. In Equation (4-1), the variables required for the wave functions are shown for the first three elements. The wave function for hydrogen requires 3 variables (r, θ, ϕ), but for helium, 6 variables are required: 3 for each electron. For lithium, 9 variables are required.

$$\text{hydrogen: } \Psi(r,\theta,\phi) \quad (4\text{-}1)$$
$$\text{helium: } \Psi(r_1,\theta_1,\phi_1,r_2,\theta_2,\phi_2)$$
$$\text{lithium: } \Psi(r_1,\theta_1,\phi_1,r_2,\theta_2,\phi_2,r_3,\theta_3,\phi_3)$$

The complexity of the solution grows for higher numbers of electrons. In fact, the Schrödinger equation can only be solved to provide closed form wave functions for single electron atoms. For two-electron atoms and higher, the Schrödinger equation must be solved numerically. There have been approximations to the Schrödinger equation which are conceptually illuminating, and we will use one approximation: the *Hartree theory*.

Hartree theory

The Hartree theory assumes Coulomb interaction between each electron and all the other electrons in the atom; however, the Coulomb potential is an approximation which accounts for the orbital of the electron. An illustration of the approach is shown in Figure 4-3 where the electron charge distribution is represented by the cloud. In the left figure, an electron is in an outer orbital of the atom, so it feels a net radial repulsion from the other electrons because the charge distribution is spatially asymmetric with respect to the electron. The electron repulsive force opposes the attractive force from the positively charged nucleus. This is referred to as electron *shielding*. Conversely, in the right side of the figure, an electron is shown in an inner orbital. In this case the electron charge distribution is nearly symmetric spatially, so the electron repulsive forces nearly cancel each other out. For an electron in an inner orbital, it feels nearly the entire nuclear attraction because there is little shielding.

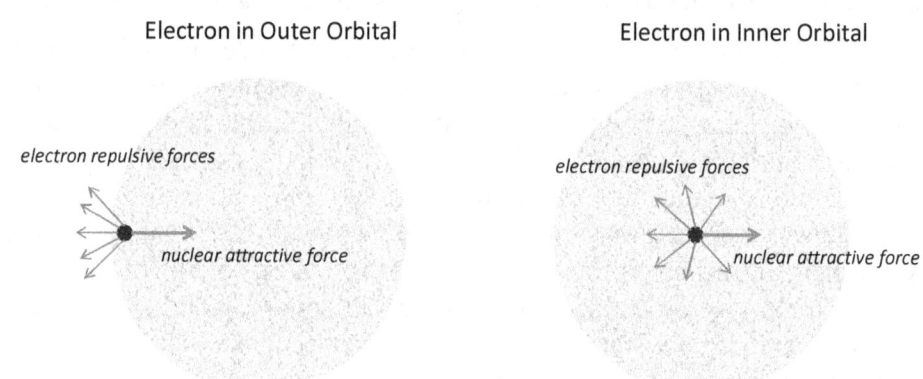

Figure 4-3: Illustration of electron shielding, and effective Coulomb potential in a multi-electron atom. An electron in an outer orbital (left) feels a net repulsion all of the other electrons in the atom. The repulsive force, represented by the vectors, reduces or *shields* the attractive force from the positively charged nucleus. For an electron in an inner orbital, the electron repulsive forces are more uniform in all directions, providing very little net repulsion.

The Hartree theory accounts for shielding by approximating an *effective Coulomb potential*, $V(r)$, for each electron in the atom. The effective potential acting on an individual electron is approximated by

$$V(r) = -\frac{Z_{n\ell}^{eff} e^2}{r} \tag{4-2}$$

which has the same form as the force for a single electron atom. For the multi-electron atom the atomic number Z has been replaced by an effective atomic number which depends upon the orbital parameterized by n and ℓ. A first guess at the values of effective potential could be

$$V(r) = \begin{cases} -\dfrac{Ze^2}{r^2} & r \to 0 \\ -\dfrac{e^2}{r^2} & r \to \infty \end{cases} \quad (4\text{-}3)$$

where r is the radial distance of the electron from the nucleus. This first guess is based upon no shielding at radial distances close to 0, so the electron feels the full force of the nuclear charge. Also, this assumes total shielding at large radial distances, so the electron feels the attractive force of the nucleus minus $Z-1$ electrons.

In practice the effective atomic number must fall between these two extremes. The example of argon follows in which an approximation of the effective atomic number depends strongly upon which shell the electron occupies, parameterized by the principle quantum number, n. Recall that $Z = 18$ for argon.

$$\text{argon} \quad \begin{aligned} Z^{eff}_{n=1} &\cong 16 \\ Z^{eff}_{n=2} &\cong 8 \\ Z^{eff}_{n=3} &\cong 3 \end{aligned} \quad (4\text{-}4)$$

It can be seen that electrons in the outer shell of argon are almost completely shielded. Z^{eff} for the second shell falls in between the values for the inner and outer shells. For the inner shell, the effective atomic number is $Z-2$ in this case, so there is very little shielding. In all multi-electron atoms, $Z^{eff} = Z-2$ is an accurate approximation for the first shell. The binding energy of the electron in an effective potential is described by Equation (4-5)

$$E_{n\ell} = -\frac{\left(Z^{eff}_{n\ell}\right)^2 R_H}{n^2} \quad (4\text{-}5)$$

where R_H is the Rydberg constant for hydrogen, and n the principal quantum number. This is the Bohr solution for the energy levels for a single electron atom which is an important result of the Hartree theory. The electrons are treated independently in order to solve for the wave functions; the electron-electron interaction is accounted for in the effective potential. Note that the electrons in outer shells have binding energy comparable to the hydrogen ground state binding energy. Using the Bohr radius formula, the mean radius of an electron is

$$\bar{r}_{n\ell} = \frac{n^2 a_0}{Z^{eff}_{n\ell}} \quad (4\text{-}6)$$

where a_0 is the Bohr radius for the hydrogen atom. It can be observed from Equations (4-5) and (4-6) that atoms with high atomic number are not very much larger than atoms with small atomic number. This is because the inner shells contract significantly.

Wave function in the Hartree theory and electron configuration notation

As described above, the key feature of the Hartree theory is to approximate the solution to the Schrödinger equation as the product of the single electron solutions. The electron-electron repulsion is not ignored, but it is accounted for in the effective Coulomb potential. For example, the helium wave function can be approximated as the product of two Ψ_{100} wave functions. As was described in detail for the single electron atom, the wave functions can also be described using the orbital notation. These two wave functions are also known as $1s$ orbitals (see below). The Pauli exclusion principle requires that these electrons have opposite spin magnetic quantum numbers.

$$\text{He}: \Psi(r_1,\theta_1,\phi_1,r_2,\theta_2,\phi_2) = \Psi(r_1,\theta_1,\phi_1) \cdot \Psi(r_2,\theta_2,\phi_2) \quad (4\text{-}7)$$

wavefunction: $\quad\quad\quad\quad\quad\quad\quad \Psi_{100+1/2} \quad\quad \Psi_{100-1/2}$

orbital $\quad\quad\quad\quad\quad\quad\quad\quad\quad\quad 1s(1) \quad\quad\quad 1s(2)$

For lithium, the wave function can be approximated as the product of three single electron wave functions. The first two electrons occupy $1s$ orbitals, while the third electron occupies a $2s$ orbital. The two $1s$ electrons must have opposite spin magnetic quantum numbers. The $2s$ electron can have either spin up or spin down ($m_s = \pm\frac{1}{2}$) since there is only one electron occupying the $2s$ orbital.

$$\text{Li}: \Psi(r_1,\theta_1,\phi_1,r_2,\theta_2,\phi_2,r_3,\theta_3,\phi_3) = \Psi(r_1,\theta_1,\phi_1) \cdot \Psi(r_2,\theta_2,\phi_2) \cdot \Psi(r_3,\theta_3,\phi_3) \quad (4\text{-}8)$$

wavefunction: $\quad\quad\quad\quad\quad\quad\quad \Psi_{100+1/2} \quad\quad \Psi_{100+1/2} \quad\quad \Psi_{200\pm1/2}$

orbital: $\quad\quad\quad\quad\quad\quad\quad\quad\quad 1s(1) \quad\quad\quad 1s(2) \quad\quad\quad 2s$

The somewhat cumbersome wave function notation can be more compactly described using the *electron configuration notation*. For example, the electron configuration notation is shown below for the first four elements:

hydrogen $\quad\quad 1s^1$
helium $\quad\quad\quad 1s^2$
lithium $\quad\quad\quad 1s^2 2s^1$
beryllium $\quad\quad 1s^2 2s^2$

The exponents are the numbers of electrons in each orbital, so for beryllium, there are 2 electrons in the $1s$ orbital, and 2 electrons in the $2s$ orbital. Identifying the spin magnetic quantum number is not necessary. In the cases where there is one electron in an orbital, the state can be either spin up or down. In the cases where there are two electrons in an orbital, they must be opposite spin. Furthermore, the magnetic quantum number (m) is not represented. For example, neon could be written as:

neon $\quad\quad 1s^2 2s^2 2p_x^2 2p_y^2 2p_z^2$; however, it is written more compactly as $1s^2 2s^2 2p^6$

There is one further abbreviation that allows electron configuration notation to be very compact. It is based on a property of the noble gases, which are elements which have the *s* and *p* orbitals for a given principal quantum number completely filled. The noble gases are represented as:

helium	$1s^2$		
neon	$1s^2 2s^2 2p^6$		
argon	$1s^2 2s^2 2p^6 3s^2 3p^6$	or	$[Ne]3s^2 3p^6$
krypton	$1s^2 2s^2 2p^6 3s^2 3p^6 4s^2\, 3d^{10} 4p^6$	or	$[Ar]4s^2\, 3d^{10} 4p^6$
radon	$1s^2 2s^2 2p^6 3s^2 3p^6 4s^2\, 3d^{10} 4p^6 4f^{14} 5d^{10} 6s^2 6p^6$	or	$[Xe]4f^{14} 5d^{10} 6s^2 6p^6$

As is seen above, for the larger atomic numbers, the lower energy orbitals are abbreviated by the noble gas electron configuration that represents them. For example, sulfur is written in electron configuration notation as $[Ne]3s^2 3p^4$. The electrons in inner orbitals making up the noble gas configuration are called *core electrons*.

Binding energies and the aufbau principle

The previous section described the electron configuration notation, and this section describes how to determine the energy levels and orbitals which are occupied by the atom's electrons. The approach is called the *aufbau principle*, not for a person's name, but from the German for "building up". The electrons in the atom fill up the available energy levels starting with the lowest, $1s$. The Pauli exclusion principle determines the number of electrons in each energy level. In a multi-electron atom, the binding energy is determined by the shell (n) and subshell (ℓ), and the order of the energy levels is shown in Figure 4-4.

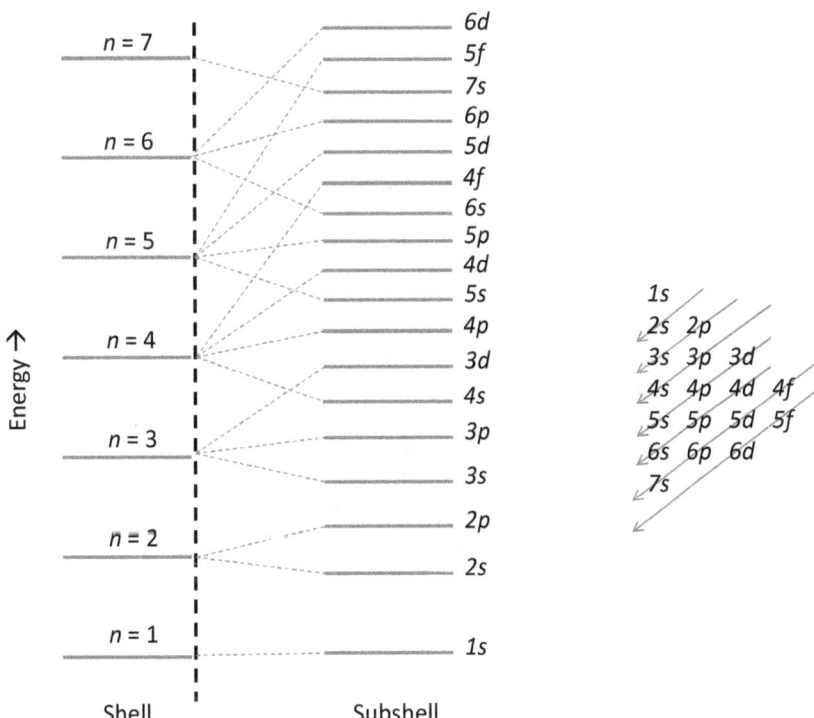

Figure 4-4: Illustration of energy levels in multiple electron atoms. The energy for an electron depends upon the shell (n) and the subshell (ℓ). On the right is an illustration of a method for determining the order of the binding energies.

The order of the binding energies for the orbitals will be described below. For now we'll assume the order of the energies in Figure 4-4 and the subshells are filled up as shown in Table 4-2. After $1s$ and $2s$ are filled, the 6 elements boron through neon have $2p$ filled. Each subshell has $2\ell+1$ orbitals, so it has capacity for $2(2\ell+1)$ electrons. Thus, $3d$ has capacity for 10 electrons. The $3d$ level has the first exceptions to the aufbau principle where both chromium and copper remove $4s$ electrons in order to have half-filled or filled $3d$. This is because filled and half-filled d shells have exceptionally lower energy.

This exception for filled and half-filled d orbitals occurs again in the filling of $4d$ which applies to the elements niobium, palladium and silver. The electron orbital filling continues up until uranium. There are other exceptions to the aufbau principle that are highlighted in the table. The binding energies have been calculated theoretically using the Schrödinger equation and the Hartree theory. The order of filling has been determined by experimental observation which validates the theoretical calculations.

The order for filling the higher levels is difficult to remember. A graphical method for determining the levels is shown in Figure 4-4. The subshells are written in order from left to right and each shell is written on a subsequent row. This is straightforward on lined paper. A diagonal arrow is drawn at about 45^0 through the subshells which results in the correct order for increasing energy.

N	Element Name	Electron Configuration	N	Element Name	Electron Configuration
1	Hydrogen (H)	$1s^1$	47	Silver (Ag)	[Kr] $5s^1 4d^{10}$
2	Helium (He)	$1s^2$	48	Cadmium (Cd)	[Kr] $5s^2 4d^{10}$
3	Lithium (Li)	$1s^2 2s^1$	49	Indium (In)	[Kr] $5s^2 4d^{10} 5p^1$
4	Beryllium (Be)	$1s^2 2s^2$	50	Tin (Sn)	[Kr] $5s^2 4d^{10} 5p^2$
5	Boron (B)	$1s^2 2s^2 2p^1$	51	Antimony (Sb)	[Kr] $5s^2 4d^{10} 5p^3$
6	Carbon (C)	$1s^2 2s^2 2p^2$	52	Tellurium (Te)	[Kr] $5s^2 4d^{10} 5p^4$
7	Nitrogen (N)	$1s^2 2s^2 2p^3$	53	Iodine (I)	[Kr] $5s^2 4d^{10} 5p^5$
8	Oxygen (O)	$1s^2 2s^2 2p^4$	54	Xenon (Xe)	[Kr] $5s^2 4d^{10} 5p^6$
9	Fluorine (F)	$1s^2 2s^2 2p^5$	55	Cesium (Cs)	[Xe]$6s^1$
10	Neon (Ne)	$1s^2 2s^2 2p^6$	56	Barium (Ba)	[Xe]$6s^2$
11	Sodium (Na)	[Ne] $3s^1$	57	Lanthanum (La)	[Xe]$6s^2$ $5d^1$
12	Magnesium (Mg)	[Ne] $3s^2$	58	Cerium (Ce)	[Xe]$6s^2 4f^1 5d^1$
13	Aluminum (Al)	[Ne] $3s^2 3p^1$	59	Praseodymium (Pr)	[Xe]$6s^2 4f^3$
14	Silicon (Si)	[Ne] $3s^2 3p^2$	60	Neodymium (Nd)	[Xe]$6s^2 4f^4$
15	Phosphorus (P)	[Ne] $3s^2 3p^3$	61	Promethium (Pm)	[Xe]$6s^2 4f^5$
16	Sulfur (S)	[Ne] $3s^2 3p^4$	62	Samarium (Sm)	[Xe]$6s^2 4f^6$
17	Chlorine (Cl)	[Ne] $3s^2 3p^5$	63	Europium (Eu)	[Xe]$6s^2 4f^7$
18	Argon (Ar)	[Ne] $3s^2 3p^6$	64	Gadolinium (Gd)	[Xe]$6s^2 4f^7 5d^1$
19	Potassium (K)	[Ar]$4s^1$	65	Terbium (Tb)	[Xe]$6s^2 4f^9$
20	Calcium (Ca)	[Ar]$4s^2$	66	Dysprosium (Dy)	[Xe]$6s^2 4f^{10}$
21	Scandium (Sc)	[Ar]$4s^2 3d^1$	67	Holmium (Ho)	[Xe]$6s^2 4f^{11}$
22	Titanium (Ti)	[Ar]$4s^2 3d^2$	68	Erbium (Er)	[Xe]$6s^2 4f^{12}$
23	Vanadium (V)	[Ar]$4s^2 3d^3$	69	Thulium (Tm)	[Xe]$6s^2 4f^{13}$
24	Chromium (Cr)	[Ar]$4s^1 3d^5$	70	Ytterbium (Yb)	[Xe]$6s^2 4f^{14}$
25	Manganese (Mn)	[Ar]$4s^2 3d^5$	71	Lutetium (Lu)	[Xe]$6s^2 4f^{14} 5d^1$
26	Iron (Fe)	[Ar]$4s^2 3d^6$	72	Hafnium (Hf)	[Xe]$6s^2 4f^{14} 5d^2$
27	Cobalt (Co)	[Ar]$4s^2 3d^7$	73	Tantalum (Ta)	[Xe]$6s^2 4f^{14} 5d^3$
28	Nickel (Ni)	[Ar]$4s^2 3d^8$	74	Tungsten (W)	[Xe]$6s^2 4f^{14} 5d^4$
29	Copper (Cu)	[Ar]$4s^1 3d^{10}$	75	Rhenium (Re)	[Xe]$6s^2 4f^{14} 5d^5$
30	Zinc (Zn)	[Ar]$4s^2 3d^{10}$	76	Osmium (Os)	[Xe]$6s^2 4f^{14} 5d^6$
31	Gallium (Ga)	[Ar]$4s^2 3d^{10} 4p^1$	77	Iridium (Ir)	[Xe]$6s^2 4f^{14} 5d^7$
32	Germanium (Ge)	[Ar]$4s^2 3d^{10} 4p^2$	78	Platinum (Pt)	[Xe]$6s^1 4f^{14} 5d^9$
33	Arsenic (As)	[Ar]$4s^2 3d^{10} 4p^3$	79	Gold (Au)	[Xe]$6s^1 4f^{14} 5d^{10}$
34	Selenium (Se)	[Ar]$4s^2 3d^{10} 4p^4$	80	Mercury (Hg)	[Xe] $6s^2 4f^{14} 5d^{10}$
35	Bromine (Br)	[Ar]$4s^2 3d^{10} 4p^5$	81	Thallium (Tl)	[Xe] $6s^2 4f^{14} 5d^{10} 6p^1$
36	Krypton (Kr)	[Ar]$4s^2 3d^{10} 4p^6$	82	Lead (Pb)	[Xe] $6s^2 4f^{14} 5d^{10} 6p^2$
37	Rubidium (Rb)	[Kr]$5s^1$	83	Bismuth (Bi)	[Xe] $6s^2 4f^{14} 5d^{10} 6p^3$
38	Strontium (Sr)	[Kr] $5s^2$	84	Polonium (Po)	[Xe] $6s^2 4f^{14} 5d^{10} 6p^4$
39	Yttrium (Y)	[Kr] $5s^2 4d^1$	85	Astatine (At)	[Xe] $6s^2 4f^{14} 5d^{10} 6p^5$
40	Zirconium (Zr)	[Kr] $5s^2 4d^2$	86	Radon (Rn)	[Xe] $6s^2 4f^{14} 5d^{10} 6p^6$
41	Niobium (Nb)	[Kr] $5s^1 4d^4$	87	Francium (Fr)	[Rn]$7s^1$
42	Molybdenum (Mo)	[Kr] $5s^1 4d^5$	88	Radium (Ra)	[Rn]$7s^2$
43	Technetium (Tc)	[Kr] $5s^2 4d^5$	89	Actinium (Ac)	[Rn]$7s^2$ $6d^1$
44	Ruthenium (Ru)	[Kr] $5s^1 4d^7$	90	Thorium (Th)	[Rn]$7s^2$ $6d^2$
45	Rhodium (Rh)	[Kr] $5s^1 4d^8$	91	Protactinium (Pa)	[Rn]$7s^2 5f^2 6d^1$
46	Palladium (Pd)	[Kr] $4d^{10}$	92	Uranium (U)	[Rn]$7s^2 5f^3 6d^1$

Table 4-2: The electron configuration notation for each of the naturally occurring elements. The subshells are filled in order using the aufbau principle. The highlighted entries mark exceptions to the aufbau principle.

Hund's Rule

In addition to the Pauli exclusion principle, there is another general rule describing how multi-electron atoms are built up. It is called *Hund's rule* and is described as follows: *the electron configuration with the highest spin magnetism has the lowest energy.* As the multi-electron atom is built up, the electrons must first occupy the $m_s = +\frac{1}{2}$ states (spin up) first because this would be a lower energy configuration.

	$1s^2$	$2s^2$	$2p_x^2$	$2p_y^2$	$2p_z^2$
H $1s^1$	↑	—	—	—	—
He $1s^2$	↓↑	—	—	—	—
Li $1s^2 2s^1$	↓↑	↑	—	—	—
Be $1s^2 2s^2$	↓↑	↓↑	—	—	—
B $1s^2 2s^2 2p^1$	↓↑	↓↑	↑	—	—
C $1s^2 2s^2 2p^2$	↓↑	↓↑	↑	↑	—
N $1s^2 2s^2 2p^3$	↓↑	↓↑	↑	↑	↑
O $1s^2 2s^2 2p^4$	↓↑	↓↑	↓↑	↑	↑
F $1s^2 2s^2 2p^5$	↓↑	↓↑	↓↑	↓↑	↑
Ne $1s^2 2s^2 2p^6$	↓↑	↓↑	↓↑	↓↑	↓↑

Figure 4-5: Diagram representing the electron states for the first ten atoms. The electrons occupy each subshell in order of energy. In partially filled subshells, the electrons fill all the $m_s = +\frac{1}{2}$ states first.

The electron configurations for the first ten atoms is shown in Figure 4-5. In each subshell that is fully occupied, there are equal numbers of spin up and spin down electrons. In the case of partially filled subshells, the available states are first filled with spin up electrons, and spin down ($m_s = -\frac{1}{2}$) electrons occupy the states that already have one electron. The highest total spin configuration is favored because it has the lowest total energy.

Single Electron and Multi-electron Atom Differences and Similarities

The first important similarity is the shape and structure of the orbitals. The nodal structure of orbitals is identical whether they occur in a multi-electron or single electron atom. Recall that there are $n-\ell-1$ radial nodes for an orbital. The functional shape for orbitals is similar as well. For example, the hydrogen atom 2s and 2p orbitals are shown in Figure 3-7. The shape of 2s and 2p orbitals for multi-electron atoms will be similar, but the size smaller because of the larger effective charge from the nucleus.

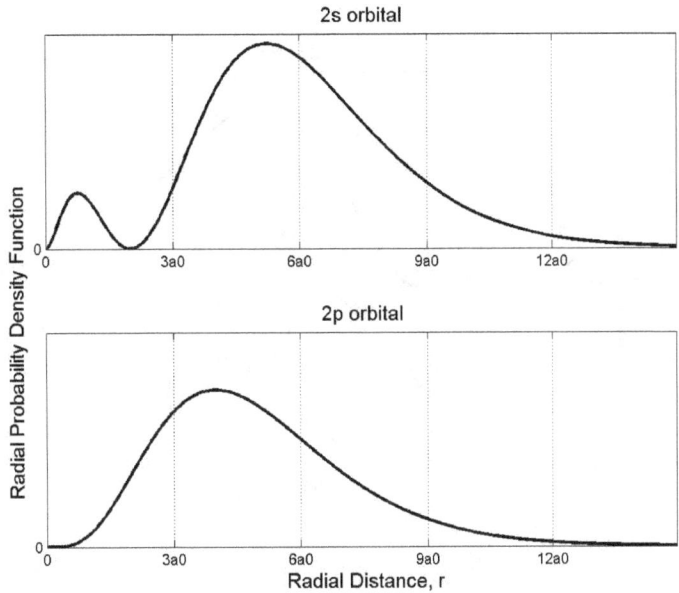

Figure 4-6: Radial probability density functions for the 2s and 2p orbitals of the hydrogen atom. Note the higher density for the 2s orbital inside the node at approximately a0.

An important difference to address is why binding energy depends upon subshell. For the single electron atom, the binding energy depended only upon the principle quantum number, n, and all orbitals with the same n are referred to as degenerate orbitals because they have equal energy. For multi-electron atoms, degeneracy is no longer true. This is due to electron shielding which was discussed above, but the particular reason that s orbitals always have lower energy than p orbitals can be illustrated using the radial probability distributions. Electrons which are less shielded, or closer to the nucleus, have a lower energy because the effective atomic number is larger. In Figure 3-7, the radial probability density is shown for the 2s and 2p orbitals. Note the higher density for the 2s orbital inside the node. The 2s electron penetrates closer to the nucleus in a probabilistic sense, and in a multi-electron atom this will serve to shield 2s electrons less than 2p electrons. The 2s orbital will be lower energy than 2p because 2s is less shielded.

Generally, s orbitals are always lower than p orbitals. A plot of the radial probability density for the 3s and 3p orbitals are shown in Figure 3-8. Note the closer penetration to the nucleus for the 3s orbital. There is a small increase in the probability density inside the first node for the 3s orbital which serves to make the 3s electron less shielded than the 3p electron.

Generally, p orbitals are always lower than d orbitals. A plot of the radial probability density (RPD) for the 3p and 3d orbitals is shown in Figure 4-8. The differences in the RPD are similar to those of the 2s and 2p orbitals. Note the closer penetration to the nucleus for the 3p orbital. There is a small increase in the probability density inside the first node for the 3p orbital which serves to make the 3p electron less shielded than the 3d electron. This suffices to make the 3p binding energy lower than the binding energy of the 3d electron.

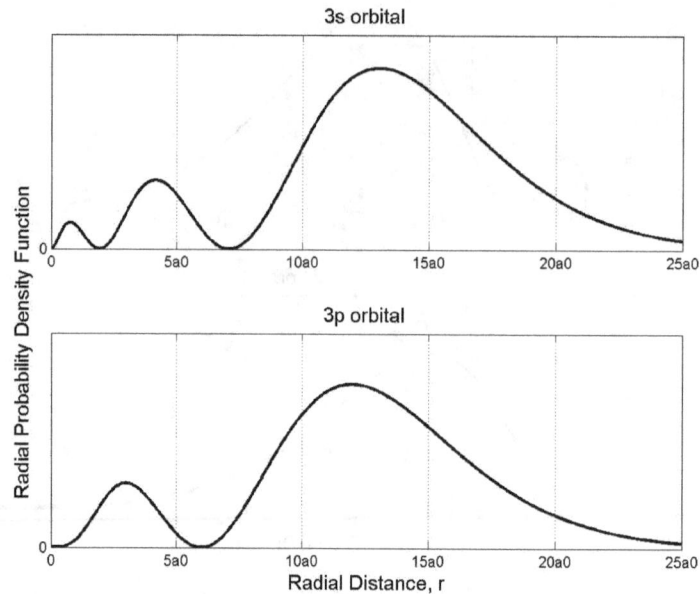

Figure 4-7: Radial probability density functions for the 3*s* and 3*p* orbitals of the hydrogen atom. The 3*s* orbital has 2 nodes and the 3*p* orbital has one node.

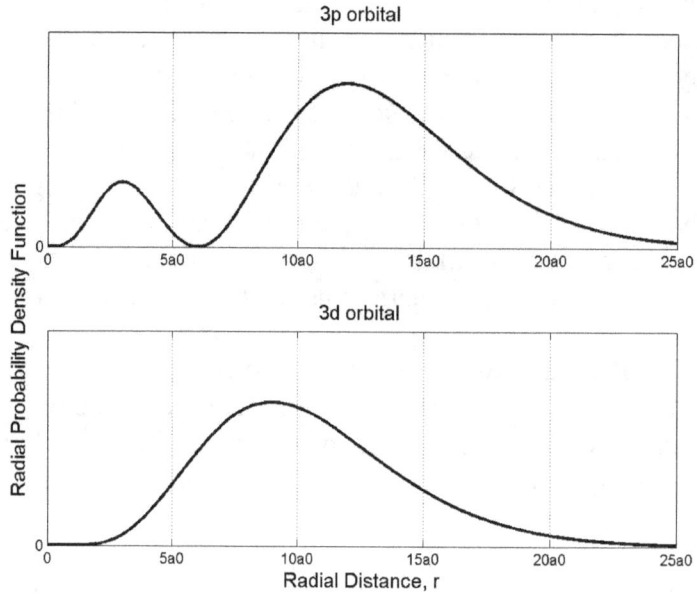

Figure 4-8: Radial probability density functions for the 3*d* and 3*p* orbitals of the hydrogen atom. The 3*d* orbital has no nodes and the 3*p* orbital has one node.

Periodic behavior and the periodic table

The Russian chemist, Dmitri Mendeleev first published his *periodic table* in 1869. The modern version of the periodic table is illustrated in Figure 4-9. Mendeleev noted the chemical properties of elements are periodic. In the periodic table the elements are ordered from left to right in a row with increasing atomic number, Z, until the chemical properties repeat. Then another row is started. This leads to columns of elements having similar chemical properties, and the columns are referred to as *groups*. The rows in the table are referred to as *periods*.

The modern way of interpreting the periodic table is due to the chemical properties of elements being determined by the outer electrons. The electrons in the outer most orbitals are referred to as the *valence electrons*, and *the valence electrons are involved in chemical reactions*. For example, the elements in the first group are the alkalis (Li, Na, K,...), and they all have one valence electron in the outer *s* orbital (s^1). Another example is the noble gases (He, Ne, Ar,...) which all have fully occupied *p* orbitals, and therefore, they have fully occupied shells.

The periodic table can thus be viewed in terms of the Hartree theory of multi-electron atoms. The ordering of the elements is done by the orbital of the outer electrons. The ordering follows the energy of the outer orbital shown in Figure 4-4. These are the same energy levels that allowed us to complete the electron configurations for all the elements in Table 4-2. For example, the second period is built up by first filling the 2*s* orbital followed by the 2*p* orbital. The outermost orbital filled in a group is shown below the columns (or groups) in Figure 4-9. Another example is the third period which is built up by filling 3*s* followed by 3*p*. The same exceptions to this building up principle hold as in Table 4-2. Because of the exceptions as well as other specific differences, the chemical similarities among groups in the periodic table are to be viewed qualitatively. The periodicity of the valence electrons leads to periodic behavior of measured properties of the elements such as *ionization energy*, *atomic radius* and *electronegativity*.

Ionization Energy
Experimentally measuring the energy levels of multi-electron atoms can be done using *photoelectric spectroscopy*. The way this works is atoms are illuminated with high energy x-rays. The x-ray photon energy is significantly higher than any electron binding energy, so when an electron bound to an atom absorbs an x-ray photon, the electron will be released from the atom. Use of high energy x-rays guarantees every electron will be ejected from any orbital. The difference in energy between the binding energy and photon energy will be the electron's kinetic energy as it travels away from the parent atom. The atom is said to be *ionized* when it loses an electron. This process is represented as an equation for neon:

$$h\nu + Ne \longrightarrow Ne^+ + e^- \tag{4-9}$$
$$h\nu + 1s^2 2s^2 2p^6 \longrightarrow 1s^2 2s^2 2p^5 + e^-$$

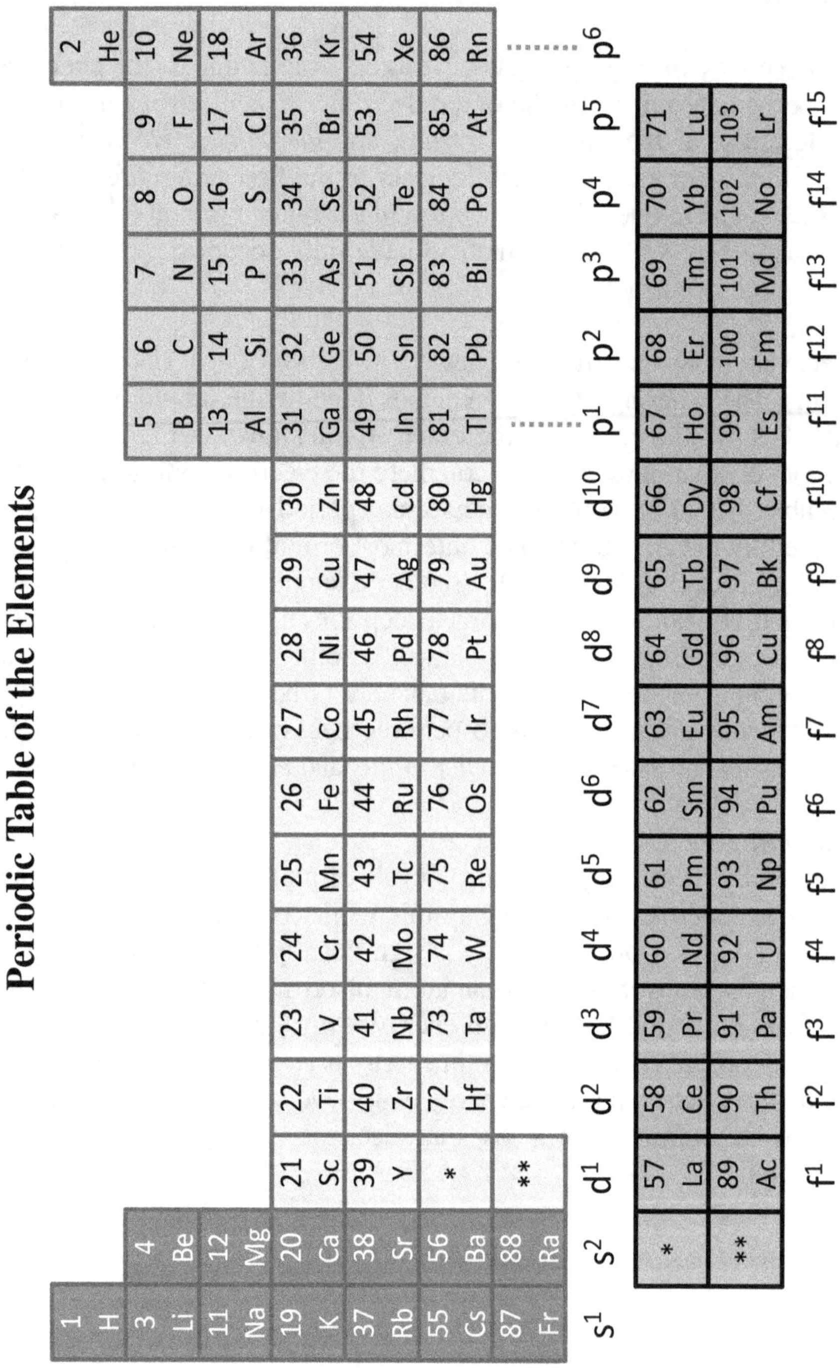

Periodic Table of the Elements

s¹	s²																
1 H																	2 He
3 Li	4 Be											5 B	6 C	7 N	8 O	9 F	10 Ne
11 Na	12 Mg											13 Al	14 Si	15 P	16 S	17 Cl	18 Ar
19 K	20 Ca	21 Sc	22 Ti	23 V	24 Cr	25 Mn	26 Fe	27 Co	28 Ni	29 Cu	30 Zn	31 Ga	32 Ge	33 As	34 Se	35 Br	36 Kr
37 Rb	38 Sr	39 Y	40 Zr	41 Nb	42 Mo	43 Tc	44 Ru	45 Rh	46 Pd	47 Ag	48 Cd	49 In	50 Sn	51 Sb	52 Te	53 I	54 Xe
55 Cs	56 Ba	*	72 Hf	73 Ta	74 W	75 Re	76 Os	77 Ir	78 Pt	79 Au	80 Hg	81 Tl	82 Pb	83 Bi	84 Po	85 At	86 Rn
87 Fr	88 Ra	**															
		d^1	d^2	d^3	d^4	d^5	d^6	d^7	d^8	d^9	d^{10}	p^1	p^2	p^3	p^4	p^5	p^6

*	57 La	58 Ce	59 Pr	60 Nd	61 Pm	62 Sm	63 Eu	64 Gd	65 Tb	66 Dy	67 Ho	68 Er	69 Tm	70 Yb	71 Lu
**	89 Ac	90 Th	91 Pa	92 U	93 Np	94 Pu	95 Am	96 Cu	97 Bk	98 Cf	99 Es	100 Fm	101 Md	102 No	103 Lr
	f^1	f^2	f^3	f^4	f^5	f^6	f^7	f^8	f^9	f^{10}	f^{11}	f^{12}	f^{13}	f^{14}	f^{15}

Figure 4-9: Periodic Table of the Elements. The valence electron configuration is illustrated in the text below the columns.

The electron has kinetic energy, *KE*, and the x-ray photon energy is $h\nu$, where h is Plank's constant and ν is the x-ray frequency. The ionization energy for the 2p electron is represented as *IE*, and it has a positive value because it is the input energy required to remove an electron. Total energy must be conserved:

$$h\nu = IE + KE \quad (4\text{-}10)$$

therefore :

$$IE = h\nu - KE$$

We return to the example of neon: neon gas is irradiated with x-rays of energy 7500 eV. The units eV are the abbreviation for electron-volt units of energy. Electrons are scattered from the neon with three different values of kinetic energy equal to 7478 eV, 7451 eV and 6630 eV. Three ionization energies can be calculated:

$$IE_{2p} = (7500 - 7478)\,eV = 22\,eV \quad (4\text{-}11)$$

$$IE_{2s} = 49\,eV$$

$$IE_{1s} = 870\,eV$$

The highest kinetic energy is associated with electrons ejected requiring the lowest ionization energy. The outer orbital (2p) has the lowest *IE*, and the inner most orbital (1s) has the highest *IE*, and this is generally true for all atoms. Another property of photoelectric spectroscopy is the number of unique kinetic energies, or unique ionization energies is equal to the number of occupied orbitals.

The *first ionization energy* (IE_1) is the energy required to remove an electron in the outermost orbital which is plotted in Figure 4-10 for each of the elements up to radon (86). In the figure, IE_1 is plotted against atomic number, Z. The ionization energy oscillates significantly with each period of the periodic table.

It is clear that the binding energy for an electron in the highest filled subshell for a noble is significantly more negative than average. The alkalis (Na, K, Rb, and Cs) have a weakly bound electron in an *s* orbital. The ionization energy is smaller than average for these elements. Alkali elements are very chemically reactive because it is energetically favorable to give up the weakly bound electron and have a more stable electron configuration of completely filled subshells. Thus for a given period, the ionization energy is highest for the noble gas and lowest for the alkalis. For example, in the second period, Li has the lowest *IE* and Ne has maximum *IE*.

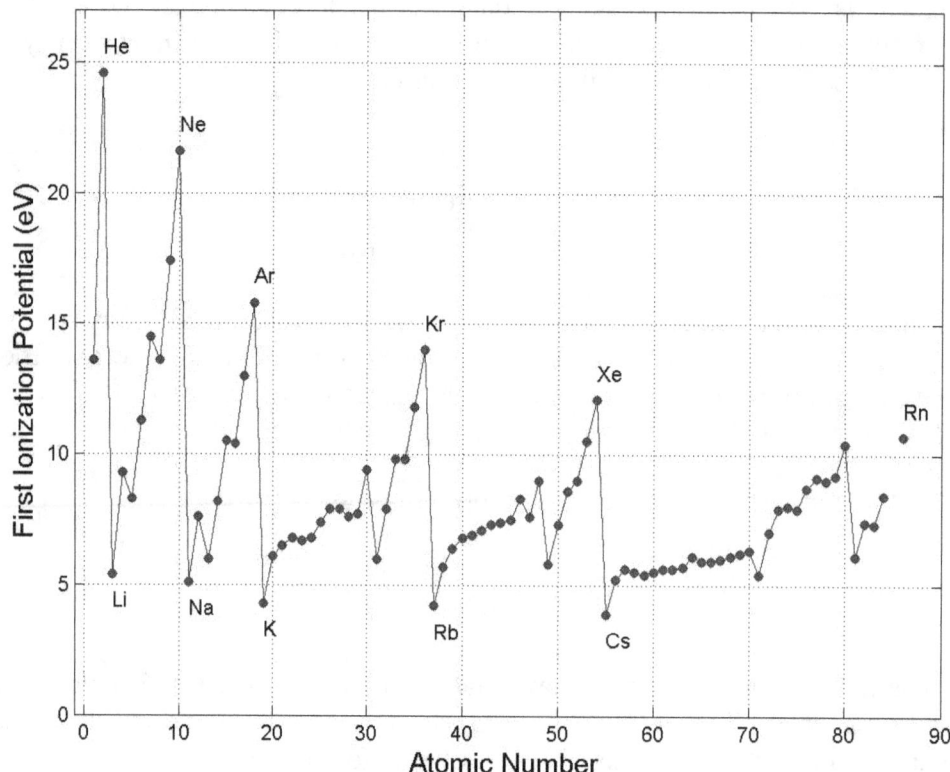

Figure 4-10: The first ionization potential plotted against atomic number. The data are from Sienko and Plane.

In the fourth period of the periodic table, the elements Sc (Z=21) through Ni (28) are filling the 3d subshell, and they have similar chemical properties and nearly the same *IE*. These elements are called the first *transition group*. Similar transition groups occur with the filling of the 4d and 5d subshells.

Atomic radius

As illustrated in the discussion on quantum mechanics, the radius of an atom is not an exact concept. The wave function of the outer electrons determine the size of the atom, but the electron distribution is probabilistic. One method of determining the atomic radius is to experimentally measure the internuclear distance of two atoms forming a bond in a crystal or molecule. The sum of the two radii of these atoms forming the bond is equal to this distance, so the atomic radius is equal to one half the internuclear distance. In one study of over 1200 bonds in all types of crystals and molecules (Slater, 1964), the measurements were shown to agree very well with a quantum mechanical calculation of radius: the calculated maximum in the radial probability density for all electrons.

Experimental atomic radii are plotted against atomic number in Figure 4-11. The periodic trend is very clear. For a given period, the atomic radius is largest for the alkalis and smallest for the halogens. Since the noble gases do not react with other elements to form bonds, the radii of the noble gases cannot be determined experimentally.

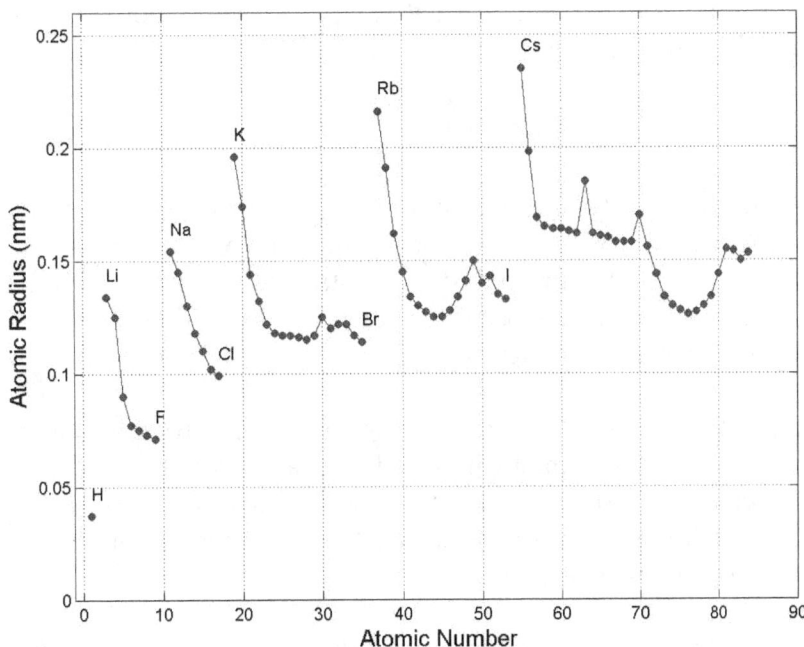

Figure 4-11: The atomic radius plotted against atomic number. The data are from Sienko and Plane.

The trend across a period of larger to smaller radius can be understood using the concept of effective potential discussed earlier. As electrons are added to a subshell, Z is increasing, and the effective potential for the inner electrons increases as well. This has the effect of contracting the inner shells. Going across a period, the outer electron is added to the same shell. Thus the inner shells contract as Z increases, and the additional electrons are added to the same outer shell and the total radius shrinks across a period. Similarly to ionization energy, the transition groups in a given period have nearly equal radii.

Electronegativity and electron affinity

The concept of *electronegativity* was first described by the American chemist, Linus Pauling in 1932. The electronegativity is always represented by χ, the Greek letter chi, and the quantity represents the strength with which an atom will attract an electron to itself. It is a calculated quantity, not a directly measured quantity, and several different methods exist for calculating electronegativity.

Shortly after Pauling's publication, the chemist Robert Mulliken proposed an alternative calculation which we will describe here because it is more intuitive. Mulliken proposed that the electronegativity for an atom is the average of the ionization energy and the *electron affinity*:

$$\chi = (IE + E_{EA})/2 \qquad (4\text{-}12)$$

where E_{EA} is the symbol for electron affinity.

The electron affinity of an atom is the amount of energy released when an electron is added to the neutral atom forming a negative ion. Using chlorine as an example, the equation describing this reaction is:

$$Cl + e^- \longrightarrow Cl^- + E_{EA} \qquad (4\text{-}13)$$

Chlorine atoms strongly attract electrons as do all of the halogens. The halogens have p^5 valence electrons and it is energetically more favorable to have a filled p subshell and have a negative charge compared to the neutral state. So E_{EA} is larger than average for the halogens, and it is positive because energy is released in the reaction. When energy is released in a reaction, it is referred to as an *exothermic reaction*.

If the value of E_{EA} is negative, input energy is required to attach an electron. When energy is consumed in a reaction, it is referred to as an *endothermic reaction*. Noble gases and alkali elements resist acquiring electrons, so they have E_{EA} values very close to zero. Electron affinity can be positive or negative depending upon the element. This is unlike ionization energy which is always positive.

Now that electron affinity has been described, the Mulliken definition of electronegativity (χ) as the average of ionization energy and electron affinity can be more easily understood. The concept is important for understanding an atom's ability to attract an electron which is shared between atoms in a molecule. In a molecule consisting of two atoms A and B, the preference of the electron for A or B depends upon how much energy is required to extract an electron from one atom (the ionization potential) and how much energy is produced when the electron binds to the other atom (electron affinity).

The calculated values of electronegativity are shown in Figure 4-12 plotted against atomic number. Given the description above that the electronegativity is the strength that an atom will attract an electron to itself, the results are consistent with other periodic properties. The highest values of χ for a given period are the noble gases. As Z increases across a period, the electrons are bound more tightly to the atom because nuclear charge increases and the electrons are added to the existing shell. At the beginning of a period, the alkalis (Li, Na, K, Rb, and Cs) have a weakly bound electron in an s orbital, and these elements have the lowest values of χ for a given period. Since electronegativity indicates a relative tendency to acquire electrons, the groups (columns) in the periodic table with low electronegativity should react with the groups with high electronegativity. By this logic, the halogens, which have very high values of χ, should form ionic bonds with the alkalis, which have the lowest values of χ.

The previous discussion of periodic properties of the elements has focused on the similarities of groups (columns) in the periodic table. In addition, there are general trends within a group in the periodic table. Generally, moving down within a group to higher values of Z, the valence electrons are less tightly bound to the atom. This is because the outer orbitals occupied by the valence electrons are further from the nucleus. Because of this effect of less binding energy of the valence electrons when moving down a column, the trends are decreasing ionization energy, increasing atomic radius, and decreasing electronegativity.

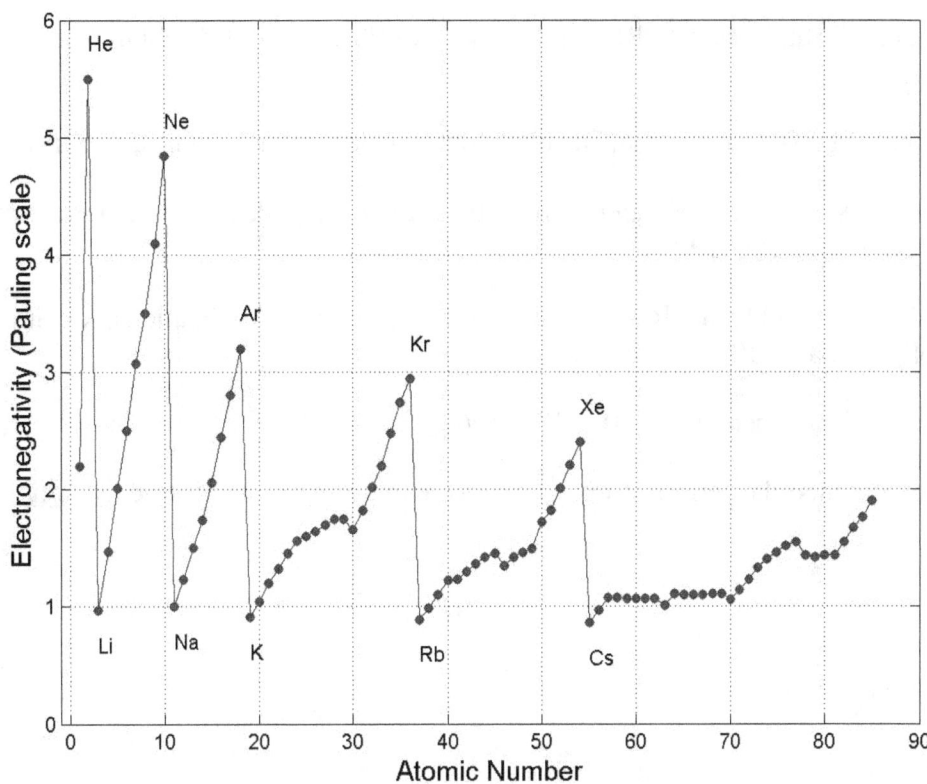

Figure 4-12: The electronegativity plotted against atomic number. The data are from WebElements.

Pauling developed the concept of electronegativity in 1932 to account for the bonding properties of molecules, and it remains an important property for determining the molecular structure of compounds, which is a subject which will be taken up in the next chapter.

References

Eisner, R. and Resnick, R., Quantum Physics of Atoms, Molecules, Solids, Nuclei, and Particles, John Wiley & Sons, 1985.

Halliday, D. and Resnick, R., Fundamentals of Physics, 2nd Ed., John Wiley & Sons, 1981.

Jammer, M., The Conceptual Development of Quantum Mechanics, McGraw-Hill, 1966.

Reger, D.L., Goode, S.R., and Mercer, E.E., Chemistry: Principles and Practice, Saunders College Publishing, 1997.

Sienko, M.J. and Plane, R.A., Chemistry, Principles and Applications, McGraw-Hill Book Company, 1979.

Slater, J.C., J. Chem. Phys. **41**, 3199 (1964).

WebElements: the periodic table on the web. http://www.webelements.com/

Chapter 5 : Chemical Bonds

Introduction

In this chapter, the *chemical bond* is introduced. In order for two atoms to become bonded together, it is required that the net energy of the molecule is lower than the energy of the two individual atoms. Many possible configurations can fulfill this requirement, and different types of bonds between neighboring atoms can be formed. The types of bonds are divided into two basic groups: *covalent* bonding and *ionic* bonding. Covalent bonding involves the sharing of two or more electrons between atoms in a molecule. In contrast, ionic bonding stems from the complete transfer of one or more electrons from one atom to another within a molecule.

In the previous chapters, we described in great detail how the arrangement of electrons in atoms and molecules is governed by quantum mechanics. This is the modern view; however, we will introduce an older non-quantum mechanical description known as *Lewis structures*. This may seem like a counterproductive twist, but Lewis structures are an extremely useful method for determining the bonding structure of molecules. Quantum mechanics is mathematically more exact, but that is also a weakness. If the objective is to rapidly determine molecular structure, a shortcut method is preferred over a detailed mathematical approach which also requires significant computation requirements. It is expected that there are exceptions to Lewis structures, but that does not diminish its use.

Covalent bonds

In molecules, such as a two atom molecule, it is required that the arrangement of their nuclei and electrons are in a lower energy state than the separate atoms. The atoms are said to be bonded together, and a *covalent bond* is a description of an electron density in which the electrons are shared between the two atoms. The shared electrons are viewed as holding the molecule together because the electrons are simultaneously attracted to both nuclei. Electrons are not always shared equally between two different atoms. Due to the electronegativity of elements, one atom will attract more of the electron density than the other atom.

Diatomic molecules

The simplest example of covalent bonding is a diatomic molecule like hydrogen (H_2). The two hydrogen atoms can form a molecule sharing two electrons because each atom can accommodate another electron in its shell. In general, covalent bonds form because the *outer shells* of the individual atoms can accommodate additional electrons. The diatomic molecule is characterized by the potential energy as a function of the spacing between the two atoms. This potential energy curve is typical for any covalent bond.

Morse potential

The physicist Philip Morse developed a formula for the potential energy of a diatomic molecule. It is important to understand the so called *Morse potential* is an *empirical* formula, which means it was developed because it fits experimental data rather than derived from other theoretical descriptions. The usefulness of the formula is not diminished because it is empirical, but it does not imply new physical principles.

The potential energy for the nuclear separation, R, is

$$V(r) = De^{-2a(R-R_0)} - 2De^{-a(R-R_0)} \qquad (5\text{-}1)$$

where D, a and R_0 are constants derived experimentally. It can be seen from Equation (5-1) that the first term becomes very large as R approaches 0, which means it is the potential energy due to a repulsive force. The source of the repulsive force is the nuclear-nuclear repulsion as the two atoms are brought closer together. The second term becomes increasingly negative as R approaches 0 which means it is the potential energy due to an attractive force. The attractive force is between the two nuclei and the electrons from both atoms, which leads to a lower energy state when the two atoms are bonded. The Morse potential has a minimum energy, $-D$, at nuclear separation, R_0, and the potential energy approaches a value of 0 as r approaches ∞.

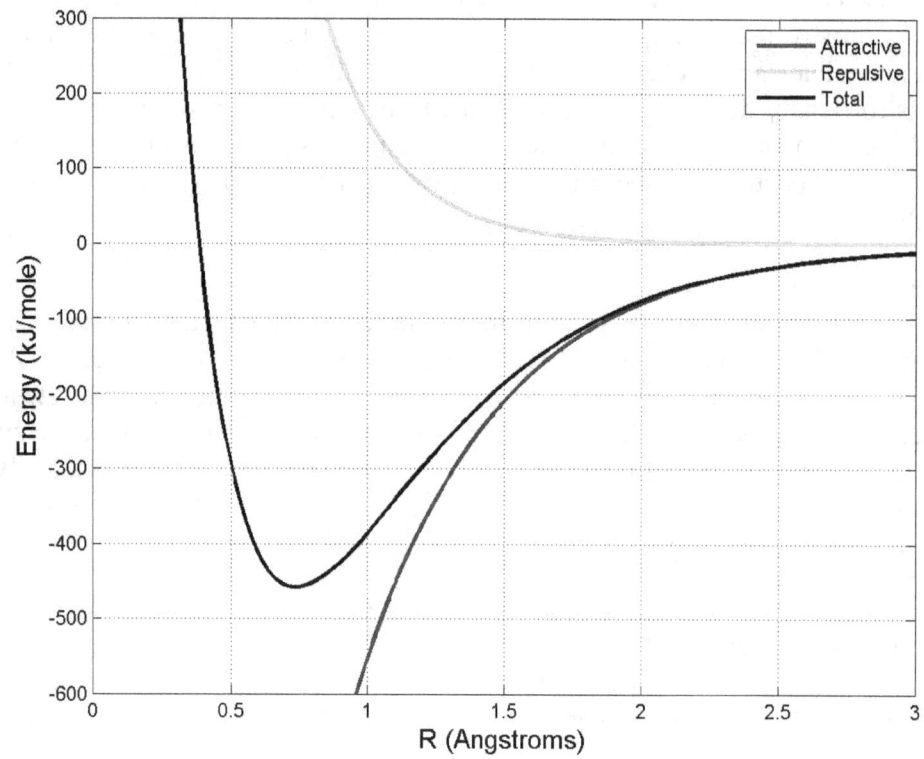

Figure 5-1: Potential energy curve (black) for the diatomic hydrogen molecule. Energy is in kJoule/mole, and the nuclear separation, R, is in Angstroms. The calculation of potential energy takes into account an attractive force (blue) and a repulsive force (green). The minimum energy occurs at the H$_2$ nuclear spacing.

The potential energy for diatomic hydrogen is plotted in Figure 5-1. The potential energy is plotted against the separation of the two hydrogen nuclei, R. Zero energy is infinite separation. The lowest energy is negative, so it is required to input energy to separate the two atoms. Since the minimum energy is negative, the atoms are bonded together. This binding energy is defines the *bond strength*. In this case bond strength is 458 kJoule/mole. The lowest energy is when the nuclear separation is 0.74 Angstrom (10^{-10} m) which is known as the *bond length*. This value of nuclear separation is twice the measured atomic radius for hydrogen, which was discussed in Chapter 4. The general characteristics of the H_2 potential energy curve are typical for any covalent bond.

Calculating the potential energy allows the comparison of bonds in other molecules. For example, the potential energy curves for diatomic hydrogen and diatomic nitrogen, N_2, are plotted in Figure 5-2. Given these two potential energy curves we can draw some comparisons between H_2 and N_2. The binding energy, or bond strength, is larger for N_2 than H_2. The N_2 bond is referred to as a stronger bond than H_2. The nuclear separation, or bond length, is also larger for N_2 than H_2. One might ask why N_2 has a longer bond than H_2. The H and N atoms are not equal size as discussed in Chapter 4 because N_2 has electrons in the *n*=2 shell and H_2 has the smaller radius *n*=1 shell populated.

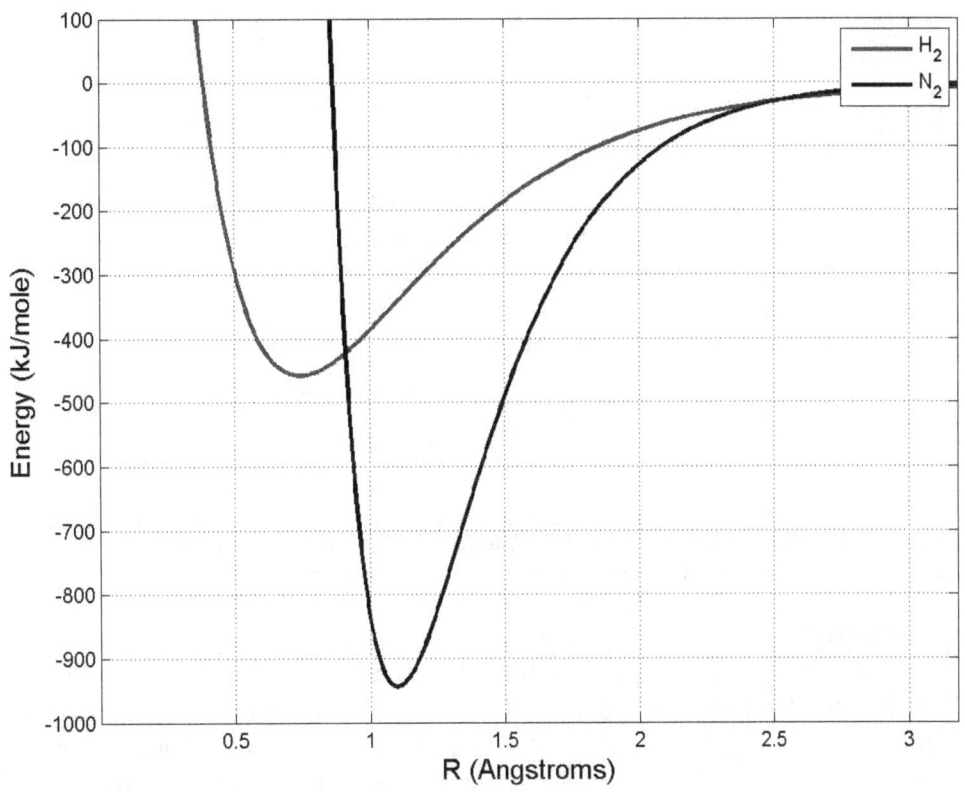

Figure 5-2: Potential energy curves for the diatomic hydrogen and diatomic nitrogen molecule. Energy is in kJoule/mole, and the nuclear separation, R, is in Angstroms. The binding energy, or *bond strength*, is larger for N_2 than H_2. The nuclear separation, or *bond length*, is also larger for N_2 than H_2.

We would like to develop some generalizations for atoms of equivalent size, so we plot the potential energy curves for diatomic oxygen, O_2, and N_2 in Figure 5-3. O and N are atomic number 8 and 7 respectively, so both atoms contain their valence electrons in the *n*=2 shell. The bond strength is larger for N_2 than O_2. The bond length for O_2 is 1.2 Angstroms as compared to 1.09 Angstroms for the N_2 bond length. *For atoms of equiva-*

lent size, stronger bonds are shorter bonds. The number of electron pairs shared in a covalent bond, which will be discussed below, can vary. The N_2 bond shares 3 electron pairs and is known as a triple bond. The O_2 molecule has an unusual structure (discussed below) and its bond shares 1 electron pair and is known as a single bond. For atoms of equivalent size, *double bonds are shorter than single bonds, and triple bonds are shorter than double bonds.*

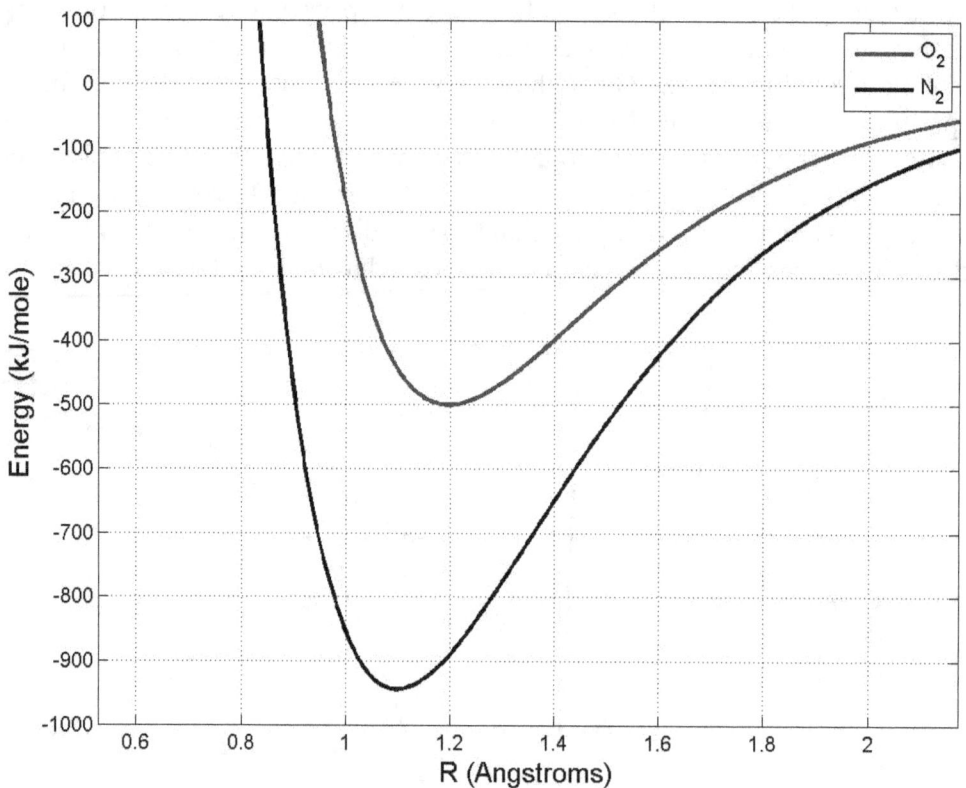

Figure 5-3: Potential energy curves for the diatomic oxygen and diatomic nitrogen molecule. Energy is in kJoule/mole, and the nuclear separation, R, is in Angstroms. The binding energy, or *bond strength*, is larger for N_2 than O_2. The nuclear separation, or *bond length*, is shorter for N_2 than O_2.

Lewis structures

The potential energy curves discussed above give insights in to how to think about covalent bonds. Specifically, we view the covalent bond as electron sharing between atoms. It is difficult to generate a potential energy curve for every atom pair in a molecule, so we require a method for representing molecules. The discussion of diatomic molecules included information that H_2 and O_2 had single bonds, and N_2 was a triple bond. One might ask how that is known and if it can be determined by a simple method. Lewis structures are such a simple method for determining covalent bonds in molecules. The method is typically taught in high school chemistry, so it certainly does not require quantum mechanics as a prerequisite.

The method was originally taught to college chemistry students by the American chemist G.N. Lewis beginning in 1902. He published the concept in a famous article, "The Atom and the Molecule", in 1916. In that article, Lewis introduced the concept of the covalent bond, the *octet rule*, and *Lewis dot structures* which will be discussed here. Lewis structures are widely used because they are quick and accurate for most cases (exceptions are discussed below). The Schrödinger equation can be difficult and time consuming, even using a numerical program, compared with the simplicity of Lewis structures.

Octet rule

When atoms share electrons in a covalent bond, all the individual atoms will have eight outer electrons. Named for the eight valence electrons, the octet rule is due to the individual atoms' tendencies to have a complete, or full, outer shell. Atoms will react so they have a valence shell electron configuration with a full octet of electrons. It is interesting that Lewis based the octet rule based on the chemical properties associate with the periodic table prior to the development of quantum mechanics. Since the development of quantum mechanics, it is known that the octet rule comes from the p and s orbitals being filled. Hydrogen is an exception to the octet rule since its outer shell is filled by two s electrons, not 8 electrons.

Periodic Table Group	Number of Valence Electrons	Valence Orbitals	Group Name	Notes
1	1	s^1	Alkali metals	H has 1 valence electron
2	2	s^2	Alkaline earth metals	
3-12	see notes	$s^2 d^{1-10}$	Transition metals	octet rule breaks down
13	3	$s^2 p^1$	Boron group	
14	4	$s^2 p^2$	Carbon group	
15	5	$s^2 p^3$	Nitrogen group	
16	6	$s^2 p^4$	Oxygen group	
17	7	$s^2 p^5$	Halogens	
18	8	$s^2 p^6$	Noble gases	He has 2 valence electrons

Table 5-1: Valence electrons associated with the Groups of the Periodic Table.

The octet rule applies to an atom's *valence electrons* which were first introduced in Chapter 4. The periodic table was viewed in terms of the quantum mechanics of multi-electron atoms. It was shown that the columns of the periodic table are related to the outer most orbitals that contain electrons. Electrons in the outer most orbitals are those with the highest energy, so these are the electrons that will be involved in chemical reactions. Table 5-1 has the valence electrons associated with the Groups of the periodic table. It can be observed that the valence electrons are those in the highest shell, i.e. the highest principle quantum number, n. The electrons in the highest s and p orbitals are the valence electrons for Group 1, 2 and 13-18.

Using the octet rule, it can be determined that the noble gases which already have a full octet form unbonded, monatomic gases. The Halogens, which have 7 valence electrons, form diatomic gases. An example, using F_2, illustrates how to draw Lewis structures. The valence electrons are drawn as dots around the individual atoms, and it is required that 2 electrons are shared between the two F atoms in order for each atom to have a full octet:

$$:\!\ddot{F}\!\cdot + \cdot\ddot{F}\!: \rightarrow :\!\ddot{F}\!:\!\ddot{F}\!:$$

From the Lewis structure, it is clear that the two F atoms share one pair of electrons. A single shared pair is defined as a single bond. We can now determine the type of bond associated with H_2 and N_2.

First, we note again that hydrogen, H, is an exception to the octet rule. Each H has one valence electron, and 2 electrons form full shell for H:

H_2 forms a single bond $H\!:\!H$

Nitrogen, N has 5 valence electrons, so the configuration that gives each N atom a full octet requires 3 shared pairs:

N_2 forms a triple bond $:\!N\!:\!:\!:\!N\!:$

So using Lewis structures we can see why H_2 and N_2 have single and triple bonds. In general, molecular structures will be more complicated, and it requires a series of formal steps to ensure that the correct Lewis structure is determined.

---

1. Draw a **skeleton structure**. H and F are always terminal atoms. The element with the lowest ionization energy goes in the middle (with some exceptions).
2. Count the total number of **valence electrons**. If there is a negative ion, add the absolute value of total charge to the count of valence electrons; if positive ion, subtract.
3. Count the **total number of electrons needed** for each atom to have a full valence shell.
4. Subtract the number in step 2 (valence electrons) from the number in step 3 (total electrons for full shells). The result is the **number of bonding electrons**.
5. Assign 2 bonding electrons to each bond.
6. If bonding electrons remain, make some double or triple bonds. In general, double bonds form only between C, N, O, and S. Triple bonds are usually restricted to C, N, and O.
7. If valence electrons remain, assign them as lone pairs, giving octets to all atoms except hydrogen.
8. Determine the **formal charge**.

Table 5-2: Formal Steps for Determining Lewis Structures (from MIT OpenCourseWare)

Lewis structure examples

The best way to learn Lewis structures is by doing examples. In the following section we will go through the formal steps for determining the structure of carbon dioxide

(CO_2), hydrogen cyanide (HCN), cyanide ion (CN^-), and the ammonium ion (NH_4^+). We will explicitly complete steps 1-7 above. For the first several examples, it is not required to calculate the formal charge. The solutions for each of the steps are self-explanatory, so the Lewis structures will be determined with minimal comment. It should be clear that determining Lewis structures is learned by doing.

Lewis structure for carbon dioxide (CO_2):

1. Draw a skeleton structure: O C O

C has lower ionization energy than O.

2. Total number of valence electrons: V = 6 (O) + 4 (C) + 6 (O) = 16

3. Total number of electrons needed: T = 8 (O) + 8 (C) + 8 (O) = 24

4. Number of bonding electrons: B = T − V = 24 − 16 = 8

5. Assign 2 bonding electrons to each bond: O:C:O

This leaves 4 bonding electrons remaining.

6. Create double or triple bonds: O::C::O

Only C has a full octet.

7. Assign remaining pairs of valence electrons: :Ö::C::Ö:

The alternative description: :Ö=C=Ö:

where the dot structures have been replaced by double bond notation. Each horizontal bar represents a bond. The remaining valence electrons are still represented by the dot notation.

Lewis structure for hydrogen cyanide (HCN):

1. Draw a skeleton structure: H C N

H is never in the middle, and C has lower ionization energy than N.

2. Total number of valence electrons: V = 1 (H) + 4 (C) + 5 (N) = 10

3. Total number of electrons needed: T = 2 (H) + 8 (C) + 8 (N) = 18

4. Number of bonding electrons: B = T – V = 18 – 10 = 8

5. Assign 2 bonding electrons to each bond: H:C:N

This leaves 4 bonding electrons remaining.

6. Create double or triple bonds: H:C:::N

H has a full shell, C has full octet.

7. Assign remaining pairs of valence electrons: H:C:::N:

This provides and alternative description: H–C≡N

where the dot structures have been replaced by single bond and triple bond notation. Each horizontal bar represents a bond. The remaining valence electrons are still represented by the dot notation.

Lewis structure for cyanide ion (CN$^-$):

1. Draw a skeleton structure: C N

This is very simple since there are only two atoms.

2. Total number of valence electrons: V = 4 (C) + 5 (N) + 1 = 10

The extra valence electron is due to the single negative charge.

3. Total number of electrons needed: T = 8 (C) + 8 (N) = 16

4. Number of bonding electrons: B = T – V = 16 – 10 = 6

5. Assign 2 bonding electrons to each bond: C:N

This leaves 4 bonding electrons remaining.

6. Create double or triple bonds: C:::N

N and C are 2 electrons short of a full octet.

7. Assign remaining pairs of valence electrons: $\left[:\!C\!:\!:\!:\!N\!: \right]^{-}$

The bracket around the structure identifies this as an ion with a single negative charge.

The compact description: $\left[:\!C \equiv N\!: \right]^{-}$

has the dot structures replaced by triple bond notation. Each horizontal bar represents a bond. The remaining valence electrons are still represented by the dot notation.

Lewis structure for the ammonium ion (NH_4^+):
1. Draw a skeleton structure

$$\begin{array}{c} H \\ H \; N \; H \\ H \end{array}$$

H atoms are always terminal atoms.

2. Total number of valence electrons: $V = 4 \cdot 1 \,(H) + 5 \,(N) - 1 = 8$

The reduction by 1 valence electron is due to the single positive charge.

3. Total number of electrons needed: $T = 4 \cdot 2 \,(H) + 8 \,(N) = 16$

4. Number of bonding electrons: $B = T - V = 16 - 8 = 8$

5. Assign 2 bonding electrons to each bond:

$$\begin{array}{c} H \\ H\!:\!N\!:\!H \\ H \end{array}$$

This leaves no bonding electrons remaining.

6. Create double or triple bonds: not applicable

7. Assign remaining pairs of valence electrons:

$$\begin{bmatrix} H \\ \overset{..}{} \\ H:N:H \\ \overset{..}{} \\ H \end{bmatrix}^+$$

The bracket around the structure identifies this as an ion with a single positive charge. The compact description:

$$\begin{bmatrix} H \\ | \\ H-N-H \\ | \\ H \end{bmatrix}^+$$

has the dot structures replaced by single bond notation. Each horizontal bar represents a bond.

Lewis structure for thionyl chloride ($SOCl_2$):
1. Draw a skeleton structure

$$\begin{array}{c} Cl \\ S \quad Cl \\ O \end{array}$$

S is in the same group as O, but S has a lower ionization energy because it is lower in the group.

2. Total number of valence electrons: $\quad V = 2 \cdot 7\ (Cl) + 6\ (S) + 6\ (O) = 26$

3. Total number of electrons needed: $\quad T = 2 \cdot 8\ (Cl) + 8\ (S) + 8\ (O) = 32$

4. Number of bonding electrons: $\quad B = T - V = 32 - 26 = 6$

5. Assign 2 bonding electrons to each bond:

There are only enough bonding electrons for 3 bonds.

6. Create double or triple bonds: not applicable

7. Assign remaining pairs of valence electrons. There are 10 lone pairs of electrons remaining:

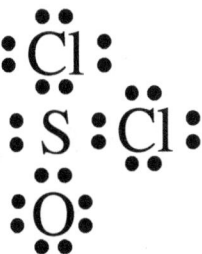

The structure for thionyl chloride has single bonds between the atoms. Clearly, all of the atoms have full octets. The total number of valence electrons is equal to the expected number, 26.

The main concept to take away from these exercises is that the electronic structure of covalently bonded molecule can be written using the rules in Table 5-2. These rules basically formalize a method for counting up the number of valence electrons in the molecule and distributing them among the atoms as lone pair electrons and shared pair electrons so that each atom acquires an octet, or noble gas configuration.

Formal Charge

The formal charge, FC, is a calculation of the quantity of the electronic charge possessed by an individual atom. Sharing of electrons is not always equal, which is indicated by electronegativity. The electronegativity represents the strength with which an atom will attract an electron to itself. Atoms which are more electronegative (higher electronegativity) attract more electron charge. Another way of thinking about FC is that it is a measure of the extent to which an individual atom has gained or lost an electron as the molecule has formed.

The formal charge computed on each atom in a molecule is defined as:

$$FC = V - L - \frac{1}{2}S \qquad (5\text{-}2)$$

where V is the number of valence electrons, L is the number of lone pair electrons, and S is the number of shared electrons. The sum of the FC for all atoms must sum to the total charge of the molecule or ion.

Some examples of how to compute the FC follow. Two examples will show how to perform the computations, and a third example illustrates the utility of FC for determining the correct Lewis structure.

Formal charge for the cyanide ion, CN⁻.

The cyanide ion, whose structure was computed above, can be succinctly written as (:C≡N:)⁻¹. In order to compute the *FC* on carbon we note that the C atom has 4 valence electrons, 6 shared electrons, and 2 lone pair electrons. The number of lone pair electrons is 2, because it's the number of electrons rather than the number of pairs that is computed. The *FC* is computed as follows:

$$FC \text{ on C:} \quad V - L - \frac{1}{2}S = 4 - 2 - 3 = -1$$

$$FC \text{ on N:} \quad V - L - \frac{1}{2}S = 5 - 2 - 3 = 0$$

The net charge is the sum of all the *FC* values over all atoms. It sums to – 1, which is the correct answer for this negatively charged ion, or *anion*.

Formal charge for the thionyl chloride, SOCl₂.

The structure for thionyl chloride, determined above is:

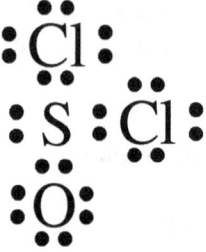

In order to compute the *FC* on oxygen we note that the O atom has 6 valence electrons, 2 shared electrons, and 6 lone pair electrons. The S atom has 6 valence electrons, 6 shared electrons and 2 lone pair electrons. Each of the Cl atoms has 7 valence electrons, 2 shared electrons and 6 lone pair electrons. The *FC* is computed as follows:

$$FC \text{ on O:} \quad V - L - \frac{1}{2}S = 6 - 6 - 1 = -1$$

$$FC \text{ on S:} \quad V - L - \frac{1}{2}S = 6 - 2 - 3 = +1$$

$$FC \text{ on Cl:} \quad V - L - \frac{1}{2}S = 7 - 6 - 1 = 0$$

The net charge, the sum of all the *FC* values over all atoms, sums to 0, which is the correct answer. The oxygen atom holds more negative charge than the sulfur atom. One might ask if this is predictable. The more electronegative atom will attract greater negative charge, so it should not be surprising that oxygen is more electronegative.

Using Formal Charge to Determine the Correct Lewis Structure

In some cases, different Lewis structures will be plausible based upon steps 1 through 7 above. These steps do not indicate a preference for two or more plausible Lewis structures. Calculating the formal charge allows us to select the correct Lewis structure from multiple plausible structures. The metric used to determine the correct structure is the sum of the absolute value of the *FC* for all atoms in the molecule. The structure with the *lowest* sum of the absolute value of *FC* is the correct structure. First, we'll do an example, and then explain this criterion.

Example of *FC* for the thiocyanate ion (CSN⁻):
For this example, we will draw all of the permutations with each of the three atoms in the center.

(::N=C=S::)⁻¹	(::C=S=N::)⁻¹	(::S=N=C::)⁻¹
FC on N = 5 – 4 – 2 = –1	5 – 4 – 2 = –1	5 – 4 – 2 = 1
FC on C = 4 – 0 – 4 = 0	4 – 4 – 2 = –2	4 – 4 – 2 = –2
FC on S = 6 – 4 – 2 = 0	6 – 0 – 4 = +2	6 – 4 – 2 = 0
Sum of |*FC*| 1	5	3

Note that the valence electrons do not change, but the number of shared electrons and lone pair electrons are variable. The correct structure is the far left structure (::N=C=S::)⁻¹ based upon the criterion of lowest sum of absolute value of *FC*. The reason this criterion works for selecting the correct structure is because the lowest potential energy structure is the more stable structure. The other two structures have greater *charge separation which leads to greater potential energy*. The *FC* criterion quantifies charge separation.

The atom with the lowest ionization energy, S, is not in the middle which is an exception to the first step in Table 5-2. One can conclude from this that Step 1 is for getting started, but not for differentiating between plausible Lewis structures. If two valid Lewis structures have the same absolute value of formal charges, *the more stable structure is the one with a negative charge on the more electronegative atom.*

Electronegativity scale of the elements

The concept of electronegativity was described in Chapter 4, and will be reviewed here. Electronegativity is represented by χ, the Greek letter chi, and the quantity represents the strength with which an atom will attract an electron to itself.

The calculated values of electronegativity are shown in Figure 5-4 plotted against period (from the periodic table) for select elements. The value of χ is the strength that an atom will attract an electron to itself, so the results are consistent with other periodic properties. The highest values of χ for a given period are the noble gases. As atomic number, Z, increases across a period, the electrons are bound more tightly to the atom. At the beginning of a period, the alkalis (Li, Na, K, Rb, and Cs) have a weakly bound electron in an *s* orbital, and these elements have the lowest values of χ for a given period. In addition, there are general trends within a group in the periodic table. Moving down within a group to higher values of Z, the valence electrons are less tightly bound to the atom, so χ decreases as one moves down a group. Electronegativity will be revisited with ionic bonding below.

Resonance

Sometime there is no simple way to describe the electronic configuration of a molecule that completely accounts for all of its properties. In this case, no single picture of the molecule is adequate, but a composite picture is required to account for its properties. This is true in the case of *resonance*, a scenario in which *no single electron configuration conforms to the measured properties of the bonds and the octet rule.*

One example of resonance is the molecule of ozone, O_3, which can have either of the following structures:

$$::O=O-O::: \quad \text{and} \quad :::O-O=O::$$

where the lone pair electrons on the center oxygen atom are suppressed in order to have a simpler representation. The two structures are equally plausible, and there is no way to prefer one over the other since the *FC* will be the same for both structures. Experimental results provide good clues. For example, the bond strengths are equal which suggests the both of these structures are incorrect somehow. In addition, bond length is an important clue. The bond length is shorter than a typical single bond length, but longer than a typical double bond length. In fact, the bonds are intermediate between a single and double bond. The bond strength is also intermediate between single bond strength and double bond strength. So the question arises: How do we draw a resonant structure? Each of the structures that conform to the octet rule is not correct, and any single structure would violate the octet rule. We represent the ozone molecule with brackets around both structures and connecting arrow to indicate that the structure is intermediate between the two:

$$(::O=O-O:::) \leftrightarrow (:::O-O=O::)$$

The representation is supposed to indicate that the molecular structure is an average of the two structures. The bonding electrons are not shared between two atoms, but rather are shared over several atoms. The structure is called a *resonance structure*, or it is equivalently called a *resonance hybrid*. The choice of the word hybrid indicates that it is a combination of the two Lewis structures.

A very similar example is the gas sulfur dioxide, SO_2, which is written as follows:

$$(::O=S-O:::) \leftrightarrow (:::O-S=O::)$$

It is similar because S has the same number of valence electrons as O. Also, experimental measurements show that the bond lengths are equal, so the structure is drawn as the resonant structure shown above.

Exceptions to the Octet Rule

There are deviations from the octet rule, and we divide these into the three categories that follow. The Lewis structures can still be drawn for these molecules in some cases, but some additional knowledge beyond the octet rule needs to be applied.

An example of a deviation is molecular oxygen, also referred to as diatomic oxygen, O_2. The Lewis structure is very simple to determine, which we will do without showing the details of the formal steps. The molecule O_2 has 12 valence electrons, and requires 16 total electrons, yielding 4 bonding electrons. The correct structure should be a double bonded structure

$$::O=O:: \quad \text{but is actually} \quad .::O\text{--}O::.$$

with a single bond and two unshared electrons, one on each O atom. This is known from experimental evidence, but cannot be predicted from Lewis structures. As was described above Lewis structures can be used for most molecules, but this is an example where Lewis structures cannot work because they do not predict the unpaired electrons in O_2. Unpaired electrons are discussed in greater detail below, but molecular orbital theory is needed to understand the actual structure of molecular oxygen.

Odd number of valence electrons

Since the octet rule works by pairing electrons, clearly an odd number of valence electrons cannot conform to the octet rule. The challenge in such scenarios is to determine where to place the single unpaired electron. Like so many Lewis structure problems, it is best to illustrate this with an example.

Example of odd number of valence electron: the methyl radical (CH_3):

A product of incomplete combustion is the *methyl radical* (CH_3). We can determine the Lewis structure using the formal rules.

1. Draw a skeleton structure

$$\begin{array}{c} H \\ H \ C \ H \end{array}$$

H atoms are always terminal atoms.

2. Total number of valence electrons: $\quad V = 3 \cdot 1 \ (H) + 4 \ (C) = 7$

There are an odd number of valence electrons.

3. Total number of electrons needed: $\quad T = 3 \cdot 2 \ (H) + 8 \ (C) = 14$

4. Number of bonding electrons: $\quad B = T - V = 14 - 7 = 7$

So, an odd number of valence electrons leads to an odd number of bonding electrons.

5. Assign 2 bonding electrons to each bond:

$$\begin{array}{c} H \\ \cdot\cdot \\ H:C:H \end{array}$$

This leaves one bonding electron remaining, and the H atoms all have full shells.

6. Create double or triple bonds: not applicable

7. Assign remaining pairs of valence electrons:

$$\begin{array}{c} H \\ \cdot\cdot \\ H:C:H \\ \cdot \end{array}$$

The extra electron is assigned as a lone electron to the C atom. The other alternatives each have physical problems. Assigning the electron as a lone electron to an H atom would give that H more than a full shell. Assigning it to a bond would make one of the bonds different than the other two bonds. The only reasonable choice is to assign the electron as unbonded to the C atom which gets C as close as is possible to a full shell (it is one electron short of a full shell).

Calculating the formal charge, *FC*, on the structure can give insight on the stability. The *FC* for this structure is computed as follows:

$$FC \text{ on H:} \quad V - L - \frac{1}{2}S = 1 - 0 - 1 = 0$$

$$FC \text{ on C:} \quad V - L - \frac{1}{2}S = 4 - 1 - 3 = 0$$

The *FC* on all of the atoms is 0, so there is not very much charge separation giving it stability. In fact, no structure could have less charge separation than this structure, so we say it has the minimum charge separation. This is the correct structure with the unpaired lone electron attached to the C atom.

This type of molecule is known as a *free radical*. A free radical is a molecule with an unpaired electron on one of the atoms. Free radicals can have positive, negative, or zero charge. Since atoms tend to form full shells, the unpaired electrons favor reactions which can lead to a full octet. Free radicals are therefore highly reactive. Not surprisingly, free radicals are important in highly reactive situations such as combustion. Since free radicals are highly reactive, they can cause damage in the body, so they are of great interest in biochemistry. For example, tobacco smoking brings combustion related free radicals into the body where they can damage DNA. Also, every advertisement for skin care products claims protection against free radicals; however, free radicals have functional roles in many biochemical processes, so they are not entirely harmful.

Example of odd number of valence electrons: nitric oxide (NO):
Another example is the gas NO. The Lewis structure is developed without detailing the formal steps. The molecule has 11 valence electrons, 16 total electrons required which leads to 5 bonding electrons. Similar to the above example, we want to avoid an odd number of bonding electrons, so it seems most reasonable to select 4 bonding electrons and assign the extra odd electron as a lone electron. This leads to 3 possible structures:

$$.::N=O: \qquad .:N=O:: \qquad :N=O::.$$

Of these three, we'll discuss the physical attributes of each. The left structure has O with 2 short of a full octet, and N with 1 extra. The right structure has N with 2 short of a full octet and O with 1 extra. The middle structure is the most likely because O has a full octet, and N is 1 short of a full octet. Calculating the formal charge, *FC*, for the middle structure (.:N=O::) follows:

$$FC \text{ on N:} \quad V - L - \frac{1}{2}S = 5 - 3 - 2 = 0$$

$$FC \text{ on O:} \quad V - L - \frac{1}{2}S = 6 - 4 - 2 = 0$$

So, the structure (.:N=O::) is correct because it has the minimum possible charge separation. Switching the lone electron to the O atom would produce a non-zero *FC* on each of the atoms in the molecule. The charge separation would increase the potential energy as compared to the correct structure. In this and the previous example the possible choices for placing the unpaired electron are narrowed down to the most likely structure using formal charge calculations.

Octet deficient molecules
Octet deficient molecules occur when the constituent atoms have too few valence electrons to conform to the octet rule. Most of these molecules are formed from Group 13 elements, the boron group. This type of exception to Lewis structures is illustrated with an example.

Example of octet deficient molecules: Boron triflouride (BF_3):
Using the Lewis structure formal rules:

1. Draw a skeleton structure

$$\begin{array}{c} F \\ F \; B \; F \end{array}$$

2. Total number of valence electrons: $V = 3 \cdot 7$ (F) $+ 3$ (B) $= 24$

3. Total number of electrons needed: $T = 3 \cdot 8$ (F) $+ 8$ (B) $= 32$

4. Number of bonding electrons: $B = T - V = 32 - 24 = 8$

So, an even number of bonding electrons needs to be assigned over 3 bonds.

5. Assign 2 bonding electrons to each bond:

$$\overset{\overset{\displaystyle F}{\cdot\cdot}}{F:B:F}$$

This leaves one bonding pair remaining, and the H atoms all have full shells.

6. Create double or triple bonds

$$\overset{\overset{\displaystyle F}{\cdot\cdot}}{F::B:F}$$

7. Assign remaining pairs of valence electrons: straightforward.

The Lewis structure rules work, but we should be suspicious that identical atomic configurations lead to single bonds in two cases and a double bond in one case. Experimental data demonstrates that all three bonds are of equal length, so molecular structure with 3 single-bonded fluorine atoms is appropriate and uses the correct number of valence electrons (24). However, this structure leaves the B atom short of an octet, or octet deficient.

$$:\!\overset{..}{\underset{..}{F}}\!:$$
$$|$$
$$:\!\overset{..}{\underset{..}{F}}\!-B-\overset{..}{\underset{..}{F}}\!:$$

We will calculate the formal charge, *FC*, for this structure as follows:

$$FC \text{ on B: } V - L - \frac{1}{2}S = 3 - 0 - 3 = 0$$

$$FC \text{ on F: } V - L - \frac{1}{2}S = 7 - 6 - 1 = 0$$

So, the structure above has the minimum possible charge separation, which in addition to the experimental information that all bonds are the same length is convincing evidence that we have determined the correct structure, which is leaves B octet deficient.

Valence Shell Expansion

The octet rule may be broken when a molecule is formed with a central atom that has a principal quantum number greater than 3 ($n>3$). In Chapter 4, it was shown that the 3*d* orbitals begin to fill after the 4*s* orbital is filled. When $n>3$, it allows for *d* orbitals to be used for shared electron pairs, which means there are more than eight shared electrons which fit around the central atom. It typically happens with a heavy atom surrounding itself with smaller atoms forming more than 4 electron pairs.

An example is phosphorous pentachloride, PCl$_5$. In the molecule the phosphorus atom is surrounded by five chlorine atoms, with each of which the phosphorus forms a covalent bond.

We are aware that if five bonds are formed around the central P in the Lewis structure, the octet rule is broken. In this case, the P atom is using five orbitals of the nine available orbitals in the $n=3$ shell. The sum of nine comes from the $3s$ (1), $3p$ (3) and $3d$ (5) orbitals.

Ionic Bonds

The previous discussion involved covalent bonds in which electron pairs are shared between the constituent atoms in a molecule. In contrast, bonds can form in which there is complete transfer of one or more electrons from one atom to another within a molecule. These bonds are referred to as *ionic bonds*, or they are sometimes called *electrovalent bonds*. Just as with covalent bonds, an ionic bond forms when the transfer of electrons and bonding of atoms results in a lower energy configuration than remaining unbonded.

Ions are formed based on their valence state; for example, an atom will lose or acquire electrons until a noble gas valence configuration is achieved. For example the alkali metals in Group 1 of the periodic chart will form a noble gas configuration by losing one electron to form an ion of charge +1. The alkaline earth elements of Group 2 lose 2 electrons forming ions of charge +2. The halogens gain an electron to form a noble gas configuration with charge -1.

The ionic charge ratios determine the molecular configuration, thus sodium chloride, NaCl, has its configuration because the charge ratios of the constituent ions. It is sometimes helpful to view the formation of ionic bonds as involving electron transfers:

$$Na + Cl \rightarrow [Na^{+1}] + [Cl^{-1}] \rightarrow NaCl$$

so the formation of NaCl can be viewed as first the electron transfer from Na to Cl, and then the electrostatic binding of the constituent atoms leading to electrically neutral molecules.

So magnesium chloride has the formula, MgCl$_2$ based on the charge ratios:

$$Mg + 2Cl \rightarrow [Mg^{+2}] + 2[Cl^{-1}] \rightarrow MgCl_2$$

Another example is magnesium nitride which has the formula Mg_3N_2 because it is composed of Mg^{+2} and N^{-3}. The ionic charge ratio is used to determine the molecular structure that yields a neutral molecule. In order to determine if a molecule contains ionic bonds, one needs to analyze the electronegativity of the constituent atoms.

The relationship between ionic bonds and electronegativity

The transition from an ionic bond to a covalent bond does not occur abruptly, but rather gradually. Like many things that have human labels applied to them, we view chemical bonds as appropriately described as ionic in some cases, and covalent in another. However, chemical bonds form a continuum from covalent to ionic. Linus Pauling quantified this concept as the *partial ionic character of a bond*. The ionic character may be estimated by using the electronegativity scale.

Electronegativity (χ) determines the strength that an atom will attract an electron to itself. The calculated values of electronegativity are shown in Figure 5-4 plotted against period for select elements. This provides a graphical representation of χ analogous to the periodic table. It can be seen that fluorine and oxygen are the most electronegative elements, hydrogen and metalloids are in the center of the scale around $\chi=2$. Metals have values of 1.7 or less.

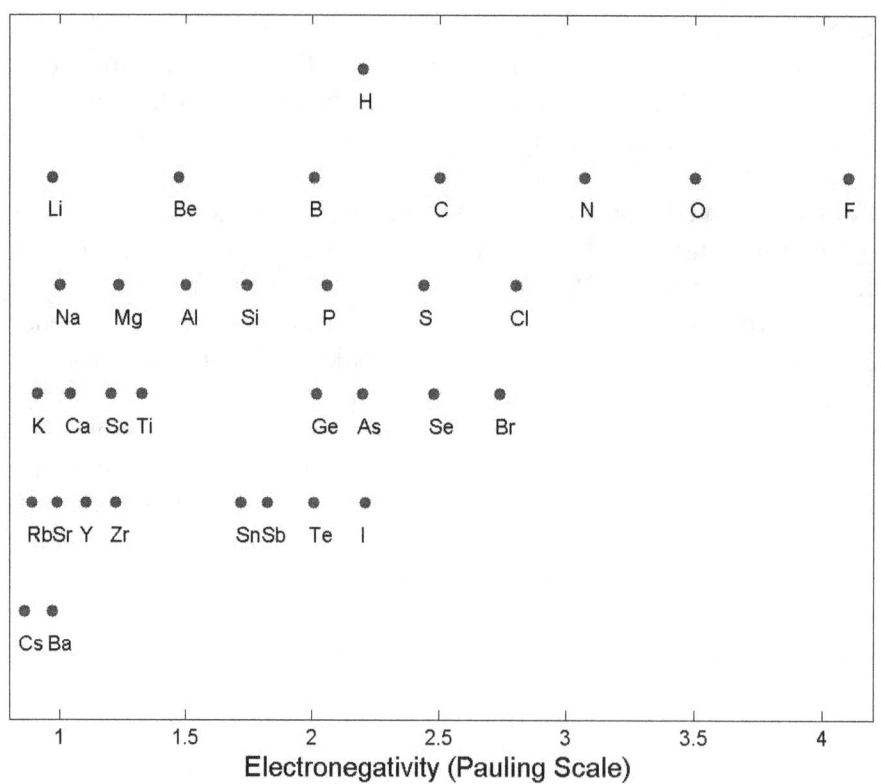

Figure 5-4: Electronegativity scale for selected elements. In descending order the values are shown for increasing period. It can be seen that the periodic trend is for electronegativity to increase as atomic number increases for a given period.

The farther two elements are from each other in χ, the greater is the amount of ionic character between them. When the difference exceeds 1.7, the bond is written as an ionic structure, otherwise it is written as a covalent structure. This rule illustrates how a continuum in nature is described with two human defined categories: covalent or ionic bond. It is important to avoid thinking about this very strictly as though something quite abrupt happens around differences of 1.7, but rather it is a convention. It is worth quoting Linus Pauling from his textbook, *College Chemistry*:

"When the separation on the scale is 1.7 the bond has about 50% ionic character. If the separation is greater than this, it would seem appropriate to write an ionic structure for the substance, and if it is less to write a covalent structure. However, it is not necessary to adhere rigorously to this rule."

Examples of several compounds that form ionic bonds are shown in Table 5-3, with their calculated separation in χ.

Name	Formula	Separation in χ
sodium fluoride	NaF	3.1
magnesium fluoride	MgF_2	2.87
lithium chloride	LiCl	1.83
calcium oxide	CaO	2.46
potassium chloride	KCl	1.89
hydrogen fluoride	HF	1.9

Table 5-3: Examples of compounds with ionic bonds and the calculated separation in χ.

In Table 5-4 are several examples of compounds that we previously determined Lewis structures for. We would expect that since they are typically treated as covalently bonded that their separation in χ would be less than 1.7. The calculations in the table show that this is the case.

Name	Formula	Separation in χ
carbon dioxide	CO_2	1.0
hydrogen cyanide	HCN	0.3 (HC) and 0.57 (CN)
ammonium ion	NH_4^+	0.87
thionyl chloride	$SOCl_2$	1.06 (S-O) & 0.36 (S-Cl)

Table 5-4: Examples of compounds with covalent bonds and the calculated separation in χ.

Energetics of ionic bonding

For elements to form a molecule, the energetics must be favorable. In other words, it is required that the net energy of the molecule is lower than the energy of the individual atoms. In the case of the formation of an ionic bind, it is viewed as a complete transfer of an electron from one atom of low ionization energy to another atom of high electron affinity. Electrostatic forces hold these ions of opposite charge together. Typical two element molecules of this type are NaF, LiCl, CaO, and KCl.

We will analyze sodium chloride, NaCl, which is common table salt. The sodium atom has low ionization energy. The ionization energy is the lowest in its period, so comparatively it is relatively easy to remove an outer electron. The chlorine atom has high electron affinity, so significant energy is released when an electron is added to its outer shell. If the sodium atom transfers an electron to a chlorine atom, then the sodium atom is positively charged and the chlorine atom is negatively charged and they can be held together by electrostatic forces. The formation of NaCl can be viewed as the following three steps:

$$Na \rightarrow [Na^{+1}] + e^-$$

$$Cl + e^- \rightarrow [Cl^{-1}]$$

$$[Na^{+1}] + [Cl^{-1}] \rightarrow NaCl$$

The net change in energy is the sum of energy released from the steps with the energy input required for the steps. This is calculated as shown below. The first step, removing an electron from sodium, requires and energy input equal to the ionization energy of Na. The second step, adding an electron to chlorine, releases energy equal to the electron affinity of Cl. The third step releases energy because the two ions are in a lower energy state when the two charged ions move close together. This energy due to electrostatic forces is calculated as follows. The energy in electron volts of two opposite charges, q_1 and q_2 held apart by a distance r is:

$$E = -1.44 \frac{q_1 q_2}{r} \qquad (5\text{-}3)$$

If the nuclear spacing of NaCl (0.28 nm), and the ionic charges (1.0 and 1.0) are plugged into Equation (5-3), the electrostatic energy is -5.1 eV. Note that the first two steps do not release sufficient energy for the bond to form, so the electrostatic force is a significant factor in calculating the net energy. In fact, electrostatic binding is more favorable for molecules in the solid state, than in the gas state. This is because electrostatic forces are active in all directions, and each negative ion will have more than one positive ion neighbor. This creates stronger attraction in the solid and liquid state than the gas state. The energy budget is shown below. The net change in energy for bonded atoms is -3.6 eV lower than for the unbonded state illustrating the stability of the bond.

$Na \rightarrow [Na^{+1}] + e^-$	$\Delta E = +5.14$ eV
$Cl + e^- \rightarrow [Cl^{-1}]$	$\Delta E = -3.61$ eV
$[Na^{+1}] + [Cl^{-1}] \rightarrow [Na^{+1}][Cl^{-1}]$	$\Delta E = -5.1$ eV
Net: $Na + CL \rightarrow [Na^{+1}] [Cl^{-1}]$	$\Delta E = -3.6$ eV

It is interesting that the energetics of electrostatic attraction is based upon classical physics, electrostatics, and does not require quantum mechanics. Quantum mechanics has been so important for modern chemistry; however, in this chapter we have made excursions into classical physics in the area of Lewis structures and electrostatic forces. We continue this excursion into classical physics in the next chapter when we discuss the kinetic theory of gases, which pre-dates the acceptance of the reality of atoms.

References

Lewis, G.N., *J. Amer. Chem. Soc.* **38**, (1916).

Morse, P.M., *Physical Review* **34**, p. 57 (1929).

Pauling, L., *College Chemistry*, W.H. Freeman and Company, 1957.

Sienko, M.J. and Plane, R.A., *Chemistry, Principles and Applications*, McGraw-Hill Book Company, 1979.

Vogel Taylor, E., 5.111 Principles of Chemical Science, Fall 2008. (Massachusetts Institute of Technology: MIT OpenCourseWare), http://ocw.mit.edu (Accessed Nov. 10, 2012). License: Creative Commons BY-NC-SA

Chapter 6 : Kinetic Theory of Gases

Introduction

Much of the early work on the kinetic theory of gases was completed before the universal acceptance of atoms. In this sense, this chapter is going back to the beginning. Chapter 1 described the microscopic analysis of atoms, meaning the discovery of the constituent parts of the atom. This led to the theory that described the mechanics of the atom: quantum mechanics (Chapter 3). There was also macroscopic analysis of the states of matter. The macroscopic properties of gases, liquids, and solids such as pressure, temperature, conductivity, heat capacity and many more properties can be predicted by the properties of the constituent atoms. In this chapter we develop the description of a collection of atoms in the gas state. The best method for understanding a gas containing a very large number of atoms is to describe the statistical properties of the constituent atoms. This topic is appropriately named *statistical mechanics*. This development of the kinetic theory in this chapter does not utilize quantum mechanics, so it is a classical physics description of gases.

Ideal gas law, equilibrium and detailed balance

Let us assume the existence of atoms. We have gone to the trouble in the previous chapters of describing the properties of atoms, molecules and chemical bonding. It would be inappropriate to describe kinetic theory as though we didn't know that the atom existed. Historically, the developers of *kinetic theory* did not have the knowledge we do today, and they were striving to demonstrate the existence of atoms. Kinetic theory as applied to the gas state treats a volume of gas as consisting of a very large number of molecules. Each molecule is in constant motion, and travels in a trajectory governed by its velocity until it collides with another gas molecule or the side of the container that holds it. The mathematical description of the mechanics of the molecules is said to be a *microscopic* description of the gas particles. These equations can then be used to make calculations for measured properties of the gas, such as the gas pressure, temperature, density, and volume. Since these measurements are based upon bulk quantities of gas, they are said to be *macroscopic* properties. Macroscopic properties were measured and known before the development of the kinetic theory. By connecting the microscopic properties of molecules to macroscopic measurements, the proponents of kinetic theory believed it would confirm the existence of atoms.

Ideal gas law

The ideal gas law is typically taught in high school chemistry. It is very simple and is extremely useful because it provides accurate predictions of gas behavior under a variety of conditions without having to worry about the chemical constituents of the gas. For a gas in a container of volume, V, held at a constant pressure P, and also held at a constant temperature T, the ideal gas law is

$$PV = nRT ,$$
(6-1)

where *n* is the number of moles of gas in the container and *R* is called the *universal gas constant*.

$$R = 8.314 \text{ J/mole-K} \tag{6-2}$$

A *mole* of atoms is a very specific number of atoms denoted by N_A which is known as Avogadro's number.

$$N_A = 6.02 \times 10^{23} \tag{6-3}$$

Alternatively, the ideal gas law can use the total number of molecules, *N*, rather than the number of moles and is written

$$PV = NkT, \tag{6-4}$$

where *k* is a different constant known as Boltzmann's constant.

Amedeo Avogadro originated the concept of the ideal gas. He asserted that a liter of gas contained the same number of entities (which turned out to be molecules) regardless of the type of gas. That was the origin of Avogadro's number, but he was forgotten for a time. Avogadro's number was given the name by the chemist Jean Perrin who earned a Nobel prize for his work which demonstrated that atoms were real. The ideal gas law connects the microscopic (number of molecules) to the macroscopic (*P*, *V*, *T*). By dividing *PV* by *T* the gas law is reduced to a formula relating only macroscopic quantities.

$$\frac{PV}{T} = \text{constant} \quad \text{(for a fixed mass of gas)} \tag{6-5}$$

This form illustrates that the ideal gas law can be viewed as an empirical law, *i.e.* a formula which fits experimental data. No assumptions about number of atoms are required to relate the three macroscopic measurements based upon experiment. However, the ideal gas law can be derived by starting with the assumption of molecules and their properties which is the basis of statistical mechanics. First, we must describe two important concepts: *equilibrium* and *detailed balance*.

Equilibrium
Equilibrium is an important concept in kinetic theory. It means that the system under study is unchanging. In order to determine that the system is unchanging, we have to also specify the *observation time*. Defining observation time is very important in chemistry. For example, imagine looking at a painting in an art museum. The paint on the canvas is dry and has been dry for a long time usually. The painting is certainly in an unchanging chemical state during the time observed. We say it is in equilibrium over this observation time. However, typically oxygen and water vapor in the air are slowly reacting with the paint and paper. The reactions are much too slow to observe during a viewing of the painting, but over an observation time of years, decades or centuries paintings can fade or flake. For example, the Sistine chapel, completed around the end of the 15th century required restoration of its painted frescoes in the late 20th century creating great controversy.

As a second example, imagine filling a rubber balloon with helium gas. The helium inside the balloon exerts a certain pressure on the balloon which is opposed by the tension of the rubber. The rubber balloon expands when filling, and the gas temperature having just expanded is usually colder than room temperature. After a few minutes the temperature of the gas inside the balloon will be at the same temperature of the air in the room. Let's assume it's for an indoor party. The temperature inside the balloon becomes equal to the air temperature because heat can be exchanged between the air and the helium through the very thin rubber. A few minutes after filling the balloon we can be certain that macroscopic properties of the helium (pressure, temperature, volume and density) will be unchanged over the next hour. So we say for such an observation time as one hour the helium filled balloon is in equilibrium.

However, as experience shows, the party balloon will shrink in volume and a few days after the party the helium will have diffused through the rubber and the balloon will sink to the floor, no longer providing any fun for children. Clearly, for observation times which are very long, we cannot assume this particular system is unchanging. So the observation time needs to be specified in order to determine if a system is in equilibrium.

At the microscopic level molecules are in constant motion. For a balloon which is approximately 1 liter in volume we can say that it contains on the order of 10^{23} molecules in constant motion. It is certainly true that the positions and velocities of each of the molecules changes constantly. The concept of equilibrium does not mean that the system is unchanging at the microscopic level. As we will discuss below, equilibrium means that the average mechanical properties, such as average molecular velocity are unchanging. In turn this means that the macroscopic properties will be unchanging.

Detailed balance

The principle of detailed balance describes the characteristics of the microscopic properties when a system is in equilibrium. A good definition of detailed balance is provided by the physicist Shang-Keng Ma:

> *"Molecular motion can be looked upon as a sequence of reactions and the reactions cause changes. The collision of the molecules causing a change of momentum is a kind of reaction. If the structure of the molecules is changed by the reaction, then this is a chemical reaction. Detailed balance implies that in equilibrium, the number of occurrences of each reaction in the forward direction is the same as that in the reverse direction. That is to say, we have equilibrium not only macroscopically, but also for each microscopic reaction."*

For example, look at the diagram representing molecules contained in a vessel in Figure 1-1. The diagram is intended to illustrate a three dimensional vessel, but only two of the three dimensions are illustrated. Each molecule has a velocity which contains a component in the x-direction, v_x. For every molecule with a velocity v_x there is a molecule with the equal and opposite value $-v_x$. In equilibrium the system is unchanging so this must hold or over time the gas would pile up on one side of the vessel. The system would get out of balance.

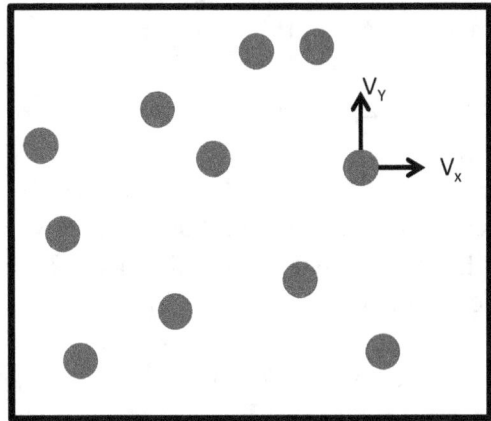

Figure 6-1: A diagram of molecules contained in a vessel. Each molecule has three velocity components v_x, v_y and v_z.

Another example is the chemical reaction

$$A + B \rightarrow C + D$$

And the reverse reaction

$$C + D \rightarrow A + B$$

where A, B, C and D represent some chemical substances. At *chemical equilibrium*, the quantities of A and B produced are unchanging, so the rates of the forward reaction exactly balance the reverse reaction. If this were not so, then the system would not be unchanging. The principle of detailed balance allows us to make statements about the distribution of velocities in a gas, and this brings us to kinetic theory.

Kinetic theory

The basic insight from kinetic theory is that macroscopic properties of a gas are caused by the motion of molecules. As we develop the kinetic properties of an ideal gas, note that chemical details are ignored. We do not treat diatomic gas molecules differently than monatomic gas molecules. The molecules are treated as tiny billiard balls in constant motion. The only molecular property needed is the molecular mass. The temperature is a measure of the kinetic energy of the molecules, and gas pressure is also a measure of the kinetic energy. We will derive this result to illustrate the theory.

Gas pressure

For a gas contained in a cube shaped vessel, such as the illustration in Figure 6-1, the molecules colliding with the walls exert a force on the wall. Let's look at the wall at right angles to the *x*-axis. In that case a molecule with velocity v_x collides with the wall and rebounds off of the wall with velocity $-v_x$ in an observation time ΔT. This is an *elastic collision* where the molecule is rebounding from the wall with the same kinetic energy and equal and opposite momentum. This assumption is a consequence of equilibrium. If a molecule gained or lost kinetic energy from each collision, then the gas would be acquiring energy from the wall or giving energy to the wall. This gas would not be in equilibrium because there would be energy flow during our observation period. For elastic collisions, the change in momentum, Δp, in this time period is:

$$\Delta p = 2mv_x \tag{6-6}$$

where m is the molecule's mass. We take all of the molecules with a velocity v_x in a volume near the wall, and these molecules will collide with the wall. In order to reach the wall in time ΔT, they would have to be within the distance $\Delta T v_x$ to reach the wall. The volume containing these molecules is then

$$volume = \Delta T v_x A \tag{6-7}$$

where A is the area of the wall. Then the number of molecules colliding with the wall in a time is ΔT

$$\# molecules = (\Delta T v_x A) \cdot \left(\frac{N}{V}\right) \tag{6-8}$$

where N divided by V is the particle density in the vessel. From Newton's laws, the force on the wall is the change in momentum per unit time

$$F = \# molecules \cdot \frac{\Delta p}{\Delta T} = A \cdot \frac{N}{V} \cdot 2mv_x^2 \tag{6-9}$$

The molecules actually have a distribution of velocities and the calculation of the force requires averaging over all of the molecular velocities. From the principle of detailed balance, we know there is a molecule with velocity $-v_x$ moving in a direction away from the wall for every molecule with velocity v_x which is colliding with the wall during the observation period. So this means that the number of molecules colliding with the wall is reduced by one half. The pressure (force per unit area) on the wall is

$$P = \frac{F}{A} = \frac{1}{2} \cdot \frac{N}{V} 2m\overline{v_x^2} = \frac{N}{V} m\overline{v_x^2} \tag{6-10}$$

where the bar over the variable means the *average of the squared velocity*. The bar over any variable is a symbol used in statistics to indicate the average of a randomly varying quantity. We have a collection of particles of identical mass, but they all have different velocities. The distribution of velocities is the familiar bell curve, a *Gaussian distribution* (discussed in more detail below). There are three velocity components, and the average of each squared velocity component is equal. The average total squared velocity is

$$\overline{v^2} = \overline{v_x^2} + \overline{v_y^2} + \overline{v_z^2} \tag{6-11}$$

So the pressure is now related to the average squared velocity of the molecules in the vessel:

$$P = \frac{1}{3}\frac{N}{V}m\overline{v^2}. \qquad (6\text{-}12)$$

Using our knowledge of the ideal gas law, we can relate the terms:

$$PV = NkT = \frac{1}{3}Nm\overline{v^2}. \qquad (6\text{-}13)$$

From Newtonian physics, the average kinetic energy of a molecule of mass m is

$$K.E. = \frac{1}{2}m\overline{v^2} \qquad (6\text{-}14)$$

It is clear that temperature is a direct measure of the kinetic energy of the moving molecules.

$$PV = NkT = \frac{2}{3}N\left(\frac{1}{2}m\overline{v^2}\right) = \frac{2}{3}U \qquad (6\text{-}15)$$

where U is the *total internal energy* of the gas. This is clear because it is the kinetic energy of each molecule multiplied by the number of molecules. Then the total energy is related to the temperature

$$U = \frac{3}{2}NkT, \qquad (6\text{-}16)$$

and the temperature can be shown to be proportional to the kinetic energy

$$T = \frac{2}{3k}\left(\frac{1}{2}m\overline{v^2}\right). \qquad (6\text{-}17)$$

Above, it was stated that molecules in a gas have a Gaussian velocity distribution. Who said the distribution is Gaussian? Two of the inventors of kinetic theory, Boltzmann and Maxwell first determined molecular velocity distributions in gases.

Maxwell-Boltzmann distribution

The Austrian physicist Ludwig Boltzmann and the Scottish physicist James Clerk Maxwell were the most important developers of kinetic theory in the 19th century. Maxwell was the first to derive the Gaussian velocity distribution for gas molecules in 1860. Boltzmann is credited with demonstrating that the physics of collisions required that gas have a Gaussian distribution of the velocities.

An example of the velocity distribution for v_x is shown in Figure 6-2. This is the probability density function for a gas of helium atoms at a temperature of 300 K. The probability density function is a measure of the frequency of finding molecules with given velocity v_x, and the shape of the distribution is the Gaussian function. The distribution is symmetric around the most probable value of v_x which is zero. Since it is symmetric, the average value, or *mean value* for v_x is zero. Why is the mean value zero? The principle of detailed balance requires that for every molecule with a velocity v_x there is a molecule with the equal and opposite value $-v_x$ for a gas at equilibrium. Even though the most probable value of v_x is zero, this does not mean there are many non-moving helium atoms since this is the motion only in one dimension. The Gaussian function also has exponential decay at large values of v_x either positive or negative. Exponential decay means the function approaches zero very rapidly with increasing v_x. This makes extremely large velocities extremely improbable.

The velocity is defined as a vector quantity in three dimensions. The magnitude of velocity, v, is referred to as the speed. From the discussion above, the important quantity to analyze is the average speed squared, so we want the distribution of molecular speeds. The distribution of speeds will allow us to calculate the quantity we want, the average kinetic energy of a molecule.

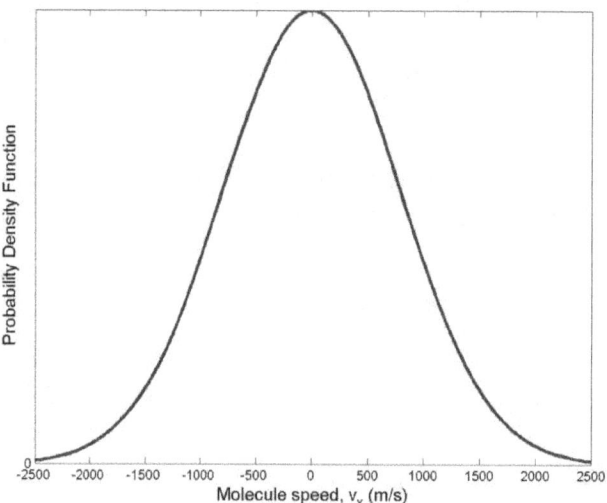

Figure 6-2: A plot of the probability distribution of helium atom velocities, v_x at a temperature of 300 degrees Kelvin.

The Maxwell-Boltzmann distribution for a gas of the same type of molecules is

$$f_v(v) = \left(\frac{m}{2\pi kT}\right)^{\frac{3}{2}} 4\pi v^2 e^{-\left(\frac{mv^2}{2kT}\right)} \tag{6-18}$$

where f_v represents the *probability density function* for the speed, v. Probability density functions were covered in Chapter 3, but are important to review here. One important insight about probability density is that it describes the frequency of finding a molecule in a given speed range. The probability density function, abbreviated PDF, $f_v(v)$ defines the probability of finding a molecule's speed between the values of v and $v + dv$. The infinitesimal derivative of v is dv. The properties of the PDF are

$$f_v(v)dv = \text{probability of speed between } v \text{ and } v + dv \qquad (6\text{-}19)$$
$$f_v(v) \geq 0$$
$$\int_0^\infty f_v(v)dv = 1$$

The first statement is a consequence of the definition. Since the PDF is a probability density in dimension v, multiplying by the infinitesimal length dv (in velocity space) gives a probability. The infinite integral follows from the meaning of probability, since the probability of finding the molecule over all speeds is 1. Since speed does not contain direction, the speed is defined as non-negative hence the integration limits from 0 to ∞. As stated above, the PDF for velocities v_x, v_y and v_z are Gaussian with mean (average) value of zero. It is worth noting that the speed distribution is not a Gaussian distribution.

The PDF for speed, v, is shown in Figure 6-3. The curves are plots of the Maxwell-Boltzmann probability distribution for molecular speed. Curves are shown for helium, diatomic oxygen and diatom nitrogen at a temperature of 300 degrees Kelvin. Once we have plotted total speed, rather than a single velocity component as in Figure 6-2, we can make some observations which are more intuitive. One observation is that the most probable speed is non-zero, and also the PDF goes to zero at a speed of zero. This is consistent with our intuition that molecules should be in constant motion. Also, note that the lighter particle, helium, has higher average speeds than the heavier molecules. Since kinetic energy is proportional to temperature, He atoms must have higher speed in order to have the same kinetic energy as the heavier atoms. Also, the PDF is exponentially decaying at high speeds: the PDF rapidly approaches zero. The PDF has two parameters, molecular mass and temperature.

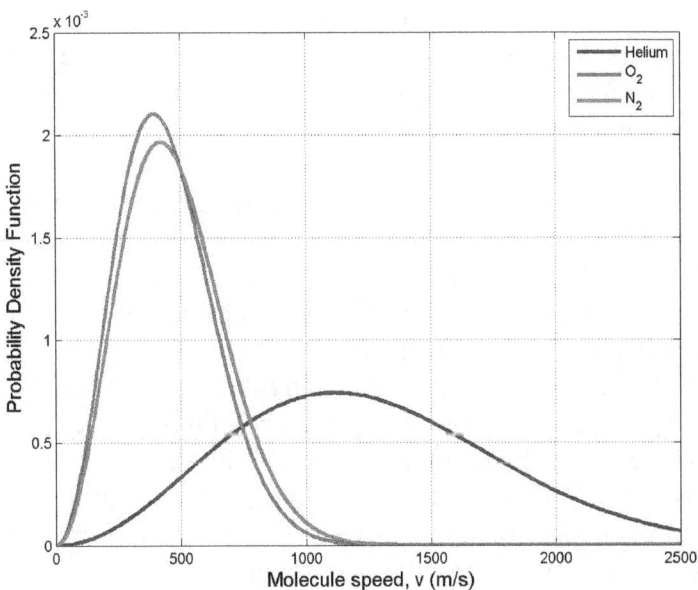

Figure 6-3: A plot of the Maxwell-Boltzmann probability distribution for molecular speed. Curves are shown for helium, diatomic oxygen and diatom nitrogen at a temperature of 300 degrees Kelvin.

From Equation (6-15) above, the temperature is proportional to the kinetic energy which means the mean squared speed is proportional to temperature:

$$T = \frac{2}{3k}\left(\frac{1}{2}m\overline{v^2}\right)$$ (6-20)

$$\overline{v^2} = \frac{3kT}{m}$$

This exactly the value of mean squared speed we get by using the Maxwell-Boltzmann PDF. The mean squared speed is defined using an integral over the PDF for speed:

$$\overline{v^2} = \int_0^\infty v^2 f_v(v)dv = \left(\frac{m}{2\pi kT}\right)^{3/2} \int_0^\infty 4\pi v^4 e^{-\left(\frac{mv^2}{2kT}\right)} dv = \frac{3kT}{m}$$ (6-21)

It is clear from Equation (6-21) that given two gases at the same temperature, the mean squared speed is higher for the lighter particles. Another way of looking at this is writing out the average kinetic energy for two different gas types at the same temperature. For oxygen and helium,

$$\frac{1}{2}m_{O_2}\overline{v_{O_2}^2} = \frac{1}{2}m_{He}\overline{v_{He}^2} = \frac{3}{2}kT$$ (6-22)

The temperature for helium is equal to the temperature for oxygen because temperature does not depend on the substance. Since the mass of helium has atomic weight 4, while diatomic oxygen is 32, the mean squared speed for helium must be 8 times higher to have the same kinetic energy and therefore same temperature. One conclusion from kinetic theory is that in a mixture of gases in equilibrium, such as air, the molecules have different speed distributions depending upon the mass.

It is clear from (6-18), that the Maxwell-Boltzmann distribution allows us to derive the ideal gas law. We could say ho-hum, this appears to be merely a re-derivation of the ideal gas law. In fact, Maxwell and Boltzmann used the correct parameters in their distribution so that it would consistent with the ideal gas law. The theory must match experimental observation. Are there any insights to be gained from kinetic theory? Let's return to the helium balloon that is mentioned in the section on equilibrium. Say we fill a rubber balloon with air and a second balloon with helium, and leave them alone at room temperature. The next day the helium balloon will be quite flat while the air-filled balloon has not noticeably changed. What's happening is the thin rubber is not an impermeable barrier to gas which can diffuse through it. The helium atoms, on average have significantly higher speed than the major constituents of air: O_2 and N_2 (see Figure 6-3). The diffusion rate is proportional to the molecular speed, so the helium balloon deflates faster.

Central limit theorem

As mentioned earlier, the velocity distributions are Gaussian, which is why the PDF for molecular speed is the form in Equation (6-18). There are very good reasons why the velocity distributions must be Gaussian in equilibrium. The Gaussian PDF comes from a mathematical theorem in statistics called the *Central Limit Theorem*. It can be derived formally, but there is a physical interpretation which is often easier to understand. We can view the velocity of a molecule in a gas as due to many interactions or collisions with other molecules in the gas. The key insight is that in a reasonably large observation time, a molecule will have a very large number of collisions with other molecules. Many of these interactions add approximately the same contribution to the molecule's velocity. The molecular velocity is therefore the sum of many equivalent contributions from the other molecules which have some velocity distribution themselves. When velocity is due to the sum of many equivalent quantities, the Central Limit Theorem requires the velocity to evolve to a Gaussian distribution even if the colliding molecules have some other distribution initially.

What is nice about this physical interpretation is that it can be easily simulated by summing a number of random variables with non-Gaussian distributions such as a uniform distribution. No matter the original distribution, after 10 to 20 summations the sum has evolved to the Gaussian. It was Boltzmann who first showed that the Gaussian distribution derived by Maxwell was due to many collisions among molecules.

Mean free path

We can calculate the rate of collisions from kinetic theory. One way to do this is to calculate the average distance a molecule travels between collisions with other moving molecules. This average distance is called the *mean free path*, and it influences chemical reaction rates in gases. Gas molecules may react chemically when they collide, so the collision rate determines the upper limit of any reaction rate.

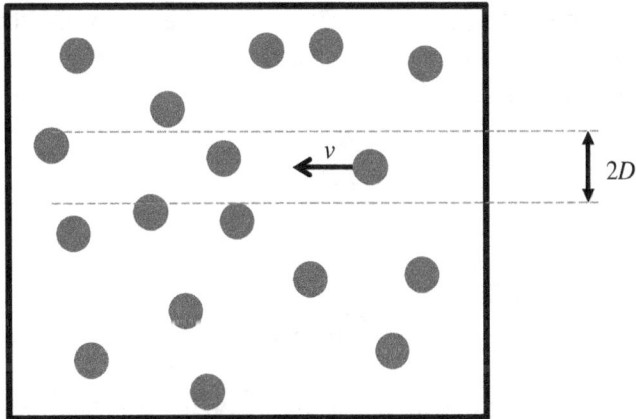

Figure 6-4: A diagram of molecules contained in a vessel. The molecule with velocity v will collide with any molecule within an imaginary tube of diameter $2D$, where D is the molecular diameter.

Consider the situation in Figure 6-4, where a molecule of diameter D traveling in a straight line between collisions has a velocity v. We envision the traveling molecule passing through an imaginary tube with diameter 2D. If another molecule has its center of mass within the tube, it will collide with the moving molecule. The traveling molecule will collide with any other molecule that is within the tube, so the number of molecules to collide with in a time, T, is determined by the volume of the imaginary tube swept out by the moving molecule. The volume of this imaginary cylinder is:

$$\text{volume} = \pi D^2 \cdot \bar{v} T \qquad (6\text{-}23)$$

Here we place a bar over the speed, v, to indicate that it is the average speed of the molecules within the vessel. The density of molecules in the vessel is the number of molecules, N, divided by the volume of the vessel, V. The number of molecules within that volume is the density of molecules multiplied by the volume of the tube.

$$\text{number of molecules} = \pi D^2 \cdot \bar{v} T \cdot \left(\frac{N}{V}\right) = \pi D^2 \cdot \bar{v} T \cdot n \qquad (6\text{-}24)$$

where n represents the molecular density. The number of molecules colliding is an average quantity, and the total path traveled is an average quantity. The average path traveled, λ, between collisions is the total path divided by the number of collisions. This path calculation is

$$\bar{\lambda} = \frac{\bar{v} T}{\pi D^2 \cdot \bar{v} T \cdot n} = \frac{1}{\pi D^2 n} \qquad (6\text{-}25)$$

Note the bar over the path to indicate it is a mean path length. This formula was derived assuming that one molecule is moving and all the other molecules are like stationary billiard balls. Of course, all the other molecules are moving too, so we need a correction to the formula assuming collisions with moving molecules. Equation (6-25) has an assumption in it that the velocity in the numerator and the denominator are the same, which they are not. The velocity in the numerator is the molecular velocity with respect to an unmoving cylinder. The velocity in the denominator is the mean relative speed with respect to other molecules. The relative velocity calculated using the Maxwell-Boltzmann distribution of molecular speeds is

$$\bar{v}_{rel} = \sqrt{2}\bar{v} \qquad (6\text{-}26)$$

so the mean free path is

$$\bar{\lambda} = \frac{1}{\sqrt{2}\pi D^2 n} \qquad (6\text{-}27)$$

The equation makes clear that the mean free path is related to the size of the molecules and the molecular density.

Collision frequency

The collision frequency will be the *average speed* of a molecule divided by the mean free path. Note that the average speed is different from the mean squared speed. We can calculate collision frequency using the mean velocity from kinetic theory

$$\bar{v} = \sqrt{\frac{8kT}{\pi m}} \tag{6-28}$$

so the collision frequency, Z, is

$$Z = \frac{\bar{v}}{\bar{\lambda}} = \sqrt{\frac{8kT}{\pi m}} \sqrt{2}\pi D^2 n \tag{6-29}$$

Note the collision frequency is related to the parameters m, D, n, and T. It should seem intuitively obvious that the collision frequency increases with the molecular diameter, D and the density, n. There are more opportunities for collision if the density is higher, and there are more opportunities to collide if the diameter is larger. Picture a room full of bouncing beach balls compared with an equal number of billiard balls bouncing around at the same density. The beach balls will collide with each other much more often. It might not be obvious that the collision rate is related to temperature and molecular mass, but this dependence comes from the calculation of average speed. Molecular speed increases with increasing gas temperature leading to higher collision rates. Also, for a given temperature, smaller mass molecules move faster.

As stated above, the collision frequency is the upper limit of any reaction rate. For mixtures of different molecular substances, the derivation of collision frequency is different. For example, consider two substances, A and B, which are in the gas state and combine chemically. In such a case we might be interested in the collision frequency to understand the reaction rate. Using a derivation from kinetic theory the collision frequency for the A and B molecules is

$$Z_{AB} = \sqrt{\frac{8kT}{\pi \mu_{AB}}} \pi r_{AB}^2 n \tag{6-30}$$

where the gas density for A and B is assumed to be the same value, n. The differences from the collision rate for the gas of a single substance are the reduced mass, μ_{AB}, replaces the mass, m. The reduced mass is derived from the masses of the A and B molecules:

$$\mu_{AB} = \frac{m_A m_B}{m_A + m_B} \tag{6-31}$$

When one molecule is significantly heavier than the other, the reduced mass is approximately equal to the smaller molecule's mass. The other quantity that is different from the single substance equation is the molecular diameter which is replaced with an effective radius of interaction, r_{AB}. This reflects and average between the cross sectional areas of the two different molecules.

Boltzmann distribution

The Maxwell-Boltzmann distribution in Equation (6-18) can be transformed to a probability density function for the kinetic energy, E. The math to perform the transformation is not difficult and is typically covered in courses on probability and statistics. The PDF for energy is

$$f_E(E) = \frac{2\pi}{(\pi kT)^{3/2}} \sqrt{E} e^{-E/kT} \tag{6-32}$$

where the only parameter is the temperature T. The particle mass is no longer a parameter. This should be expected because it is part of the kinetic energy calculation. For all types of molecules, the average kinetic energy was shown to be $3kT/2$. A plot of the PDF is shown in Figure 6-5 for three different temperatures. The shape of the PDF is very different for diverse temperatures. As temperature increases, note the PDF gets wider. Since the area under the PDF has to be 1 (Equation (6-19)), the peak value must get smaller as the PDF gets wider. The wider PDF means more high energy molecules, which is to be expected at higher temperatures.

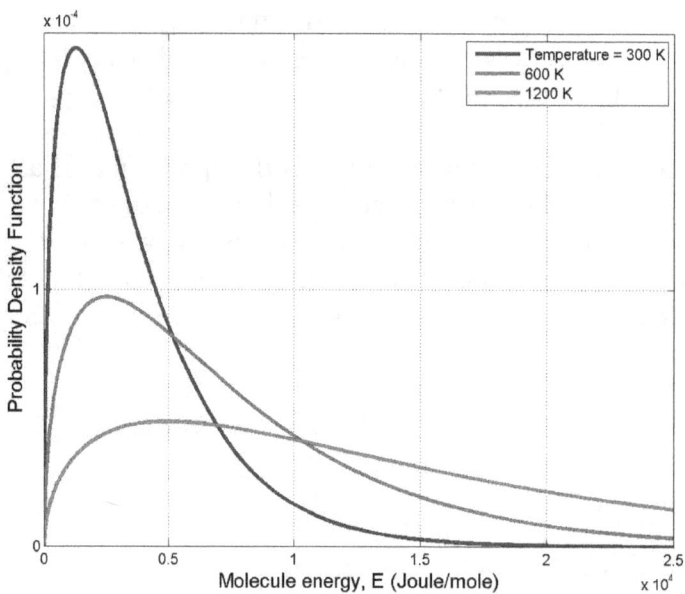

Figure 6-5: A plot of the Maxwell-Boltzmann probability distribution for molecular kinetic energy. Curves are shown for three different temperatures: 300K, 600K and 1200K.

The Maxwell-Boltzmann distribution with respect to energy is important for reactions and reaction rates. For example, consider the different temperature distributions in Figure 6-5. Let us imagine that there is a reaction that has *activation energy* $E_A = 2 \times 10^4$ Joules/mole. The activation energy is the minimum energy required for the reaction to proceed. For the 300K temperature distribution, the PDF above E_A is very nearly zero. By comparison, for the 1200K temperature distribution, the PDF at E_A is not much lower than the peak value of the PDF. There are substantial numbers of molecules with kinetic energy larger than E_A. We would expect that the reaction would go forward at a faster rate at 1200K than at 300K because the molecules have energy greater than the activation energy.

Boltzmann derived a more general probability distribution for states as a function of the energy. The probability of a molecule being in a state with energy E is proportional to an exponential term

$$P(E) \propto e^{-E/kT} \tag{6-33}$$

In the case of an ideal gas, the probability of finding a molecule with energy in a region E to $E+dE$ at the energy E is

$$P(E) = \frac{2\pi}{(\pi kT)^{3/2}} \sqrt{E} e^{-E/kT} dE = \text{constant} \cdot g(E) e^{-E/kT} dE \tag{6-34}$$

where $g(E)$ is the density of the molecules with energy E. The function $g(E)$ is called the *density of states*. Since the right hand side is a probability, the constant term can be calculated by integrating over all energies

$$P(E) = \frac{g(E)}{\int_0^\infty g(E) e^{-E/kT} dE} \cdot e^{-E/kT} dE \tag{6-35}$$

Equation (6-35) is known as the *Boltzmann distribution*. For the ideal gas, it merely re-writes the Maxell-Boltzmann distribution with density of states

$$g(E) = \sqrt{E} . \tag{6-36}$$

We will show in the next chapter that the Boltzmann distribution is more general. We will be able to use it for applications other than ideal gases, and we will be able to apply it to quantum states.

It is worth pointing out that we've discussed the kinetics of gases in this chapter and modeled the gas molecules as tiny billiard balls. We've made no assumptions that take advantage of our modern knowledge of molecular physics including quantum mechanics. Of course, these methods pre-date quantum mechanics, so that is not surprising. However, we will find as we begin to discuss properties of real gases, as opposed to ideal gases, that we cannot understand certain properties without quantum mechanics. It also is very interesting that the Boltzmann distribution extends to quantum mechanics even though Boltzmann derived it decades before quantum mechanics.

References

Cohen, L., "The history of noise," *IEEE Signal Processing*, Nov. 2005.

Cropper, W.H., *Great Physicists*, Oxford University Press, 2001.

Feynman, R., Leighton, R.B., Sands, M., *Lectures on Physics*, Adeson-Wesley Publishing Company, 1966.

Halliday, D. and Resnick, R., *Fundamentals of Physics*, 2nd Ed., John Wiley & Sons, 1981.

Ma, S., *Statistical Mechanics*, World Scientific Publishing, 1985.

Rosser, W.G.V., *An Introduction to Statistical Physics*, Ellis Horwood, 1982.

Chapter 7 : Internal Degrees of Freedom of Gases

Introduction

In the previous chapter, it probably struck the reader how modern physics was not required to understand the material. The existence of atoms was assumed, but the detailed quantum mechanical description of the atom was not required. It is interesting that very simple assumptions, such as thinking of atoms has hard elastic particles was sufficient to derive properties of gases. However, one has to have a modern quantum mechanical description of molecules when analyzing internal degrees of freedom. After all, classical mechanics is no longer valid at atomic scales. In this chapter we discuss internal dynamics, such as rotation and vibration. We focus in detail on diatomic molecules, such as N_2 and H_2. We demonstrate how to determine the heat capacity of diatomic gases by combining the Boltzmann distribution with quantum mechanical determination of the energy levels.

Internal degrees of freedom

In the last chapter, we gained many insights by modeling molecules in a gas as though they were billiard balls, but they are not that simple. Molecules can tumble around in a complicated manner depending upon the configuration of the molecules. The molecules can rotate and vibrate, so a fraction of the kinetic energy in the gas can be stored in rotational or vibrational motion. This is referred to as internal kinetic energy to distinguish it from the translational velocity in the x, y, and z directions. Since energy can be stored internally, temperature can be measured in terms of the *internal degrees of freedom* as well as the three dimensional velocities.

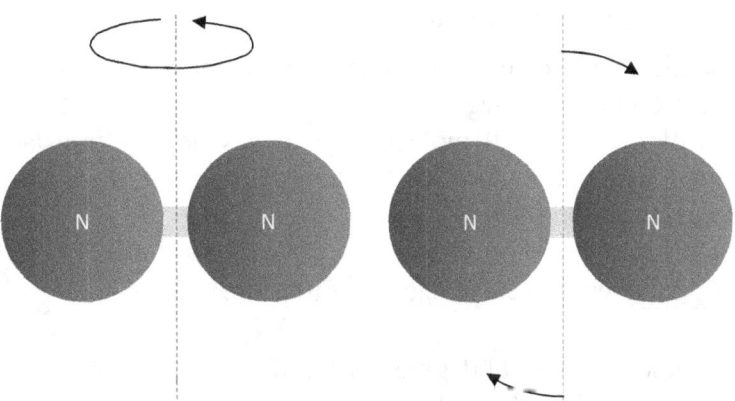

Figure 7-1: A diagram of the diatomic molecule N_2 illustrating two modes of rotation.

A good way to think about an internal degree of freedom is motion other than translational motion in three dimensions. As we know, molecules consist of atoms that are bonded together and these atoms could have kinetic energy due to *modes* of motion. The term mode is used as a synonym for degree of freedom. As an example, let's consider the diatomic nitrogen molecule, N_2. The molecule consists of two nitrogen atoms bonded together, so the molecule can rotate in space. The molecule can be visualized as a being shaped like a dumbbell, and a cartoon of the N_2 molecule is shown in Figure 1-1 as a dumbbell- shaped object rotating. There are two different axes about which N_2 can rotate, so we say it has two rotational modes. For the first mode (left), the motion is entirely in one plane, and for the second mode the motion is in a plane at right angles to the first mode. These two rotational modes are *degenerate*, which means they have equal energy levels (we'll discuss the quantum rotational energy levels below). For given rotational velocities, the energy of each of the two rotational modes is equal because the moment of inertia is equal for the two modes.

Another type of internal molecular motion is vibration. Consider the N_2 molecule which has a bond which is like a spring connecting two masses. The two masses are the two N atoms. The spring can be compressed or stretched, and this will cause the masses to vibrate about an average distance, r, which is the bond length. The displacement will be a periodic oscillation in time. A cartoon of this kind of motion is illustrated in Figure 7-2.

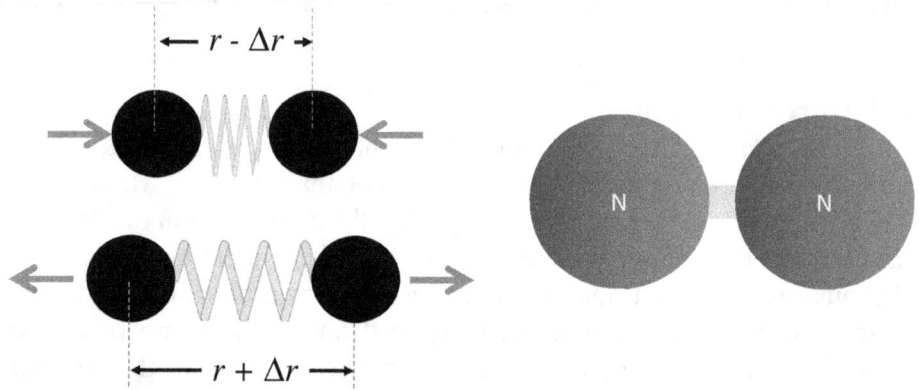

Figure 7-2: A diagram of the diatomic molecule N_2 illustrating that vibrational modes are like the compression and stretching of a spring between two masses.

The diatomic nitrogen molecule has 6 degrees of freedom. The translational motion in the x, y and z directions are 3 degrees of freedom, and there are 2 rotational, and 1 vibrational degrees of freedom. All molecules in a gas have 3 translational degrees of freedom. There is a formula to calculate the internal degrees of freedom for a molecule made up of n atoms. In general,

$$\text{number of internal degrees of freedom} = 3n - 3, \tag{7-1}$$

which correctly predicts 3 internal degrees of freedom for N_2: two degenerate rotational modes, and one vibrational mode. All diatomic molecules have two rotational and one vibrational mode.

Linear and non-linear molecules

A linear molecule lies straight along an axis that extends through all of the atomic nuclei. Examples of linear molecules include all diatomic molecules, carbon dioxide (CO_2), and hydrogen cyanide (HCN). The constituent atoms of non-linear molecules do not lie straight on a single axis. An example includes water (H_2O) which has bonds forming a 104.5 degree angle. Another example is ammonia (NH$_3$) which has three bonds forming a pyramidal three dimensional shape. A linear molecule always has only 2 rotational modes, so linear molecules have $3n$-5 vibrational modes. Non-linear molecules always have 3 rotational modes, so they have $3n$-6 vibrational modes.

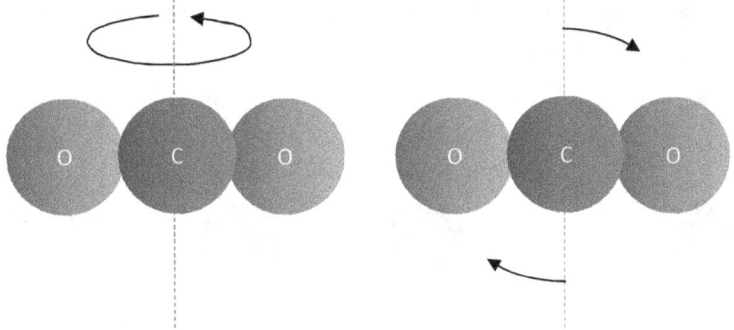

Figure 7-3: A diagram of the carbon dioxide molecule CO$_2$ illustrating two modes of rotation.

Let's consider the internal modes of the carbon dioxide molecule. The rotational modes are illustrated in Figure 7-3. The two modes have the same moment of inertia, and the same center of mass, so the rotational modes are degenerate. Since it is a linear molecule, there should be $3n$-5 = 4 vibrational modes. The vibrational behavior is more complicated than that of diatomic molecules. In the case of CO$_2$ which has three atoms with the carbon in the middle, the two bonds are identical. Two of the vibrational modes are illustrated in Figure 7-4. The symmetric stretch mode, shown on the right, has each of the bonds compressing and stretching simultaneously. At any point in time, the displacement of each of the atoms from the equilibrium position is identical. The stretching of the bond is said to be *in phase*. The anti-symmetric stretch mode, shown on the left has the displacement of the two atoms *out of phase*. The displacement has periodic behavior in time, and the anti-symmetric mode has the displacement 180 degrees out of phase. The maximum stretch of one bond occurs at a time when the maximum compression is occurring in the other bond. These two vibrational modes are not degenerate.

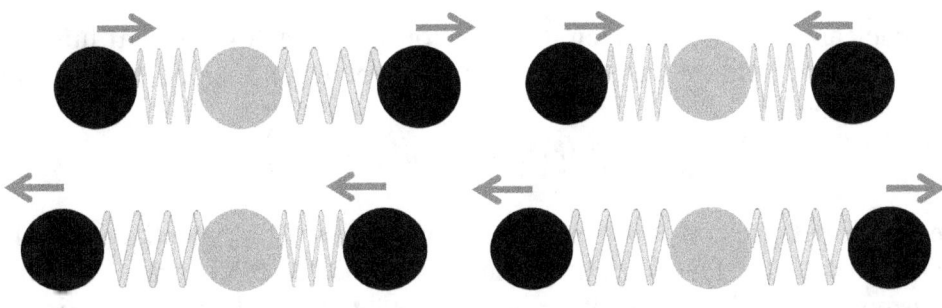

Anti-symmetric Stretch Mode   Symmetric Stretch Mode

Figure 7-4: A diagram of the symmetric and anti-symmetric stretch modes of the carbon dioxide (CO$_2$) molecule. CO$_2$ is modeled as three masses with the two bonds as springs.

The remaining two vibrational modes of CO$_2$ are known as bending modes and are illustrated in Figure 7-5. In the mass-spring model, it is clear that the springs can be displaced at right angles to the compression and stretching axis. The displacement of the atoms from their equilibrium position is then orthogonal to the stretch displacement. This bending displacement has periodic behavior in time, and as shown in Figure 7-5 the displacement of the outer O atoms is 180 degrees out of phase with the displacement of the center C atom. One bending mode has displacement of the atoms in the plane of the paper. The second bending mode has displacement in an orthogonal plane. These bending vibrational modes are degenerate.

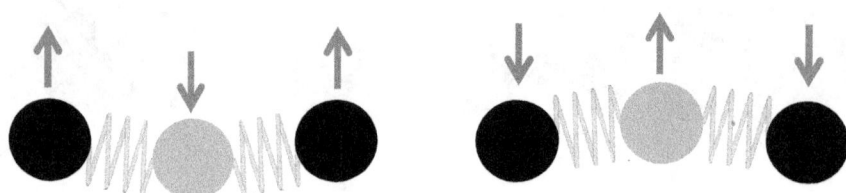

Bending Mode

Figure 7-5: A diagram of the bending modes of the carbon dioxide (CO_2) molecule. CO_2 is modeled as three masses with the two bonds as springs. One bending mode has displacement of the atoms in the plane of the paper, the second bending mode has displacement in a plane orthogonal to this plane.

Energy levels of vibrational modes

The mechanical vibration of the atoms can be analyzed to determine the allowable energy levels. As we know, the mechanics of atomic level processes need to be determined using quantum mechanics. We will use the N_2 molecule as an example. The vibrational modes of N_2 are quantized, so the molecule vibrates with only certain frequencies or energies. In order to determine the energy levels, we need to know the interaction energy of the two N atoms which was discussed in Chapter 5. The potential energy function for N_2, also called the Morse potential is

$$V(r) = De^{-2a(R-R_0)} - 2De^{-a(R-R_0)} \qquad (7\text{-}2)$$

for the nuclear separation between the two atoms, R. The exponential functions can be expanded into an infinites series. The Morse potential can then be approximated by the first terms of the series:

$$V(r) \cong D\left(-1 + a^2(R-R_0)^2\right) \qquad (7\text{-}3)$$

where D, a and R_0 are constants derived experimentally. This potential energy function is the same form as the function for two masses connected by a spring. The *mass-spring model* is not just for illustration, it is also an excellent mathematical approximation of the behavior of diatomic molecules. The form of Equation (7-3) indicates that the potential energy must have a minimum value equal to $-D$ when the nuclear spacing is at its equilibrium value R_0. For any displacement from R_0 the energy increases because the second term is squared. A plot of the potential energy for the N_2 molecule is shown in Figure 7-6 with the mass-spring approximation plotted in blue. The dissociation energy is zero, and it is clear that the mass-spring approximation deviates significantly at the higher energies. However, it is a good approximation for the lowest energies. This will prove to be acceptable because we are interested in the vibrational modes when the molecular energy is very near equilibrium. In many chemistry textbooks, the mass spring model is expanded and a spring constant is calculated. It is probably inappropriate to get too carried away with the mass-spring model. It is an appropriate model because the potential energy curve is approximated by a quadratic function like Equation (7-3), but to start calculating properties of a fictitious spring don't provide any additional insights.

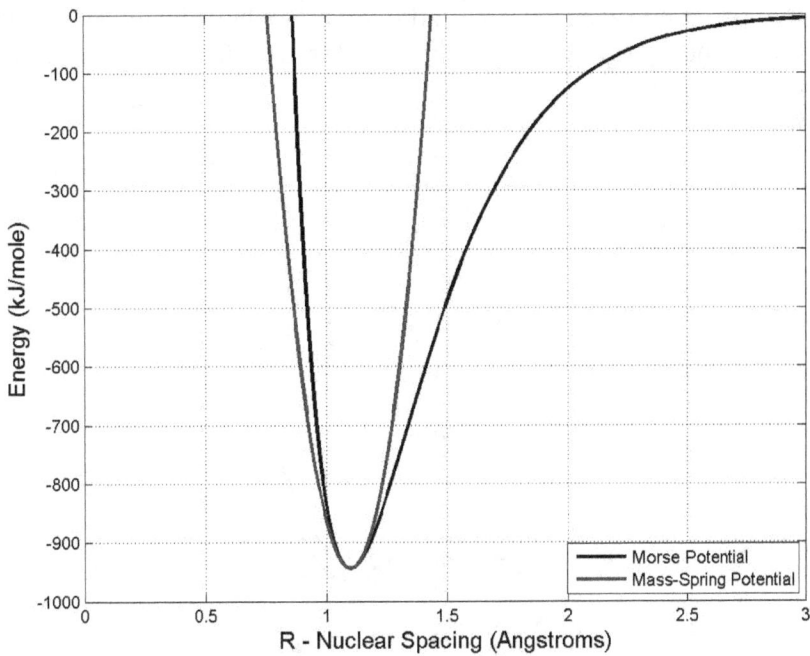

Figure 7-6: Potential energy curve for the diatomic nitrogen molecule. Energy is in kJoule/mole, and the nuclear separation, R, is in Angstroms. The blue line is the potential energy of the mass-spring model of the molecule.

A potential energy function like the form in Equation (7-3) is characteristic of an oscillating system referred to as the *harmonic oscillator*, and it is taught in introductory quantum mechanics classes. It is always taught because it is very easy to solve and the harmonic oscillator is relevant to many problems, including the dynamics of gases. The harmonic oscillator energy levels are a very good approximation of the vibrational energy levels, and the quantized energy levels for the potential in Equation (7-3) are

$$E_n = E_0 \left(n + \frac{1}{2} \right) \qquad (7-4)$$

where the index n is the quantum number of the vibrational state. Note that the lowest energy vibrational mode, when $n = 0$, is non-zero. This lowest vibrational energy, $E_0/2$, is called the *zero point energy*. Higher energies, for non-zero quantum numbers are called the first excited state ($n = 1$), the second excited state ($n = 2$), etc. Note that the energy levels are equally spaced, and the spacing is E_0. Typically, the energy levels are illustrated by plotting them on the same graph as the potential energy function which is shown in Figure 7-7. In the figure, the energy levels of the harmonic oscillator are shown as red horizontal lines. The level of the line is the energy of the quantum vibrational mode, and the extent of the line in the R dimension can be viewed as the change in spacing as the atoms oscillate back and forth periodically. The higher energy levels are associated with greater displacement in R, which is consistent with our understanding of energy in a mass spring system. If the stretch, or compression, displacement is increased the amount of potential energy in the system is increased. The potential energy is converted to kinetic energy as the atoms move. So the vibrational modes can be viewed as a mass-spring system in constant motion, and on average half the energy is potential energy, half is kinetic energy. The total energy is stored in the vibrational mode is that in Equation (7-4).

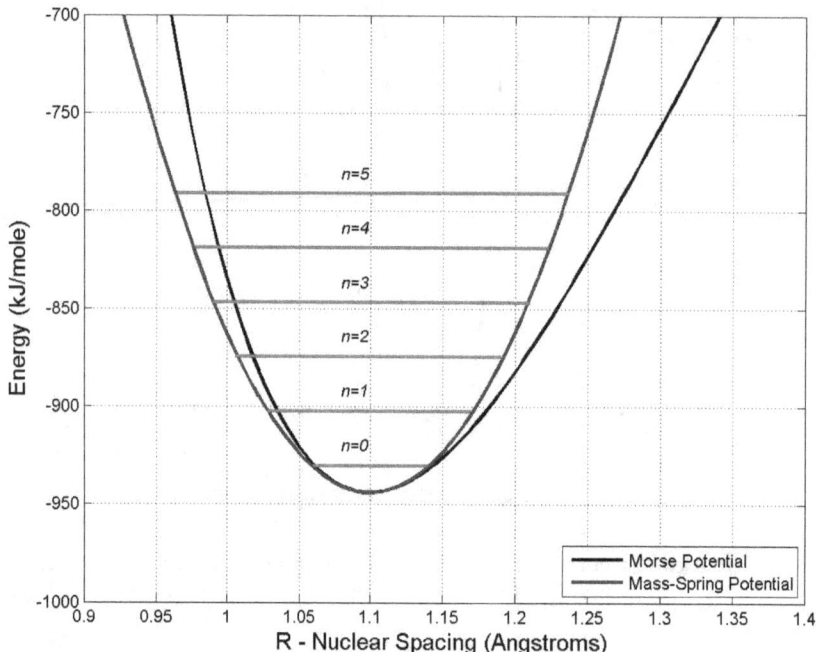

Figure 7-7: Potential energy curve for the diatomic nitrogen molecule. The blue line is the potential energy of the mass-spring model of the molecule. The red horizontal lines represent the energy levels of the first 6 quantum vibrational modes of N_2.

Vibrational spectroscopy

Since the vibrational state of diatomic has discrete energy levels, the molecule is able to change energy levels by absorption or emission of a photon. The photon will have an energy which is the difference in the vibrational energy levels. We rewrite Equation (7-4) in terms of photon energy $E=h\nu$,

$$E_n = h\nu\left(n+\frac{1}{2}\right) = hc\frac{1}{\lambda}\left(n+\frac{1}{2}\right) \qquad (7\text{-}5)$$

and it is clear that the energy differences between vibrational modes determine the photon frequency (ν) and wavelength (λ). The constant c is the speed of light. In Equation (7-5), a change in one energy level is accompanied by the emission of a photon which always has the same frequency. These vibrational modes can be investigated using *infrared spectroscopy*. The vibrational energies are low enough that photons emitted or absorbed due to vibrational mode changes are in the infrared region, which have lower energies and longer wavelengths than visible light. For infrared spectroscopy, the units for the photon frequency, v, are reported in wavenumber, which is cm^{-1}. Wavenumber in this case is exactly $1/\lambda$, which is somewhat unusual. Nevertheless, wavenumber is the traditional way to report frequency for infrared spectroscopy. Consider the various diatomic molecules in Table 7-1, where the value for E_0 and frequency are shown. Note the differences and similarities between H_2, N_2, O_2, NO and HCl. In each of these molecules, the vibrational energy levels are much smaller than the dissociation energy. The way infrared spectroscopy works is an infrared source, such as a laser, is tuned to very specific wavenumbers. The laser light is passed through a gas and the absorbed wavenumbers reflect energy level differences.

	H_2	N_2	O_2	NO	HCl
vib. energy hv (kJ/mole)	50.7	27.8	18.5	22.4	34.4
Wavenumber v (cm^{-1})	4237	2320	1549	1868	2875
dissociation energy (kJ/mole)	435	945	498	630	431
$h/8\pi^2 I$ (kJ/mole)	0.71	0.024	0.018	0.020	0.13

Table 7-1: **Vibrational and rotational energy levels for various diatomic molecules. The frequency is reported in wavenumber, calculated in cm^{-1}. Dissociation energy is from WebElements.**

Boltzmann distribution and vibrational mode population

The Boltzmann distribution can predict the fraction of molecules in each of the vibrational states. Recall from Chapter 6, the probability of finding a molecule with energy in a region E to $E+dE$ is

$$P(E) = \frac{g(E)e^{-E/kT}dE}{\int_0^\infty g(E)e^{-E/kT}dE} \qquad (7\text{-}6)$$

where $P(E)$ is the probability and $g(E)$ is the density of the molecules with energy E, also called the density of states. The application of the Boltzmann distribution to discrete energy states, means the integral is now a sum. There is one mode for each vibration energy, so the density of states $g(E) = 1$ for all modes, and the probability of a molecule with vibrational energy E_n is

$$P(E_n) = \frac{e^{-E_n/kT}}{\sum_n e^{-E_n/kT}}. \qquad (7\text{-}7)$$

It is clear from Equation (7-7) that the probability depends only on temperature, T, since the energy levels are a property of the molecule. The probability represents the fraction of the molecules in each of the energy states, so the sum of all probabilities must equal 1. A plot of the probabilities for the N_2 molecule is in Figure 7-8. Note that the probabilities are plotted against the log of the temperature in degrees Kelvin. The Kelvin scale begins at 0, but we begin the plot at 100 °K. Room temperature is approximately 300 °K. The temperature scale in the figure goes up to 10,000 °K. At room temperature and below most molecules are in the ground vibrational state, $P(n = 0) = 0.99999$. As the temperature increases, a fraction of the molecules are raised into excited states. For example, at 1000 °K, 96% of the molecules are in the ground state, $P(n = 0) = 0.96$, and most of the molecules in excited states are in the first excited state, $n =1$. As the temperature increases from 1000 to 10,000 °K, the fraction of molecules in the ground state drops 30%, $P(n = 0) = 0.3$, and significant fractions populate the first 4 excited states. As the temperature approaches even higher temperatures, the fraction of molecules in each of the states will approach the same value. This is called equipartition at high temperatures.

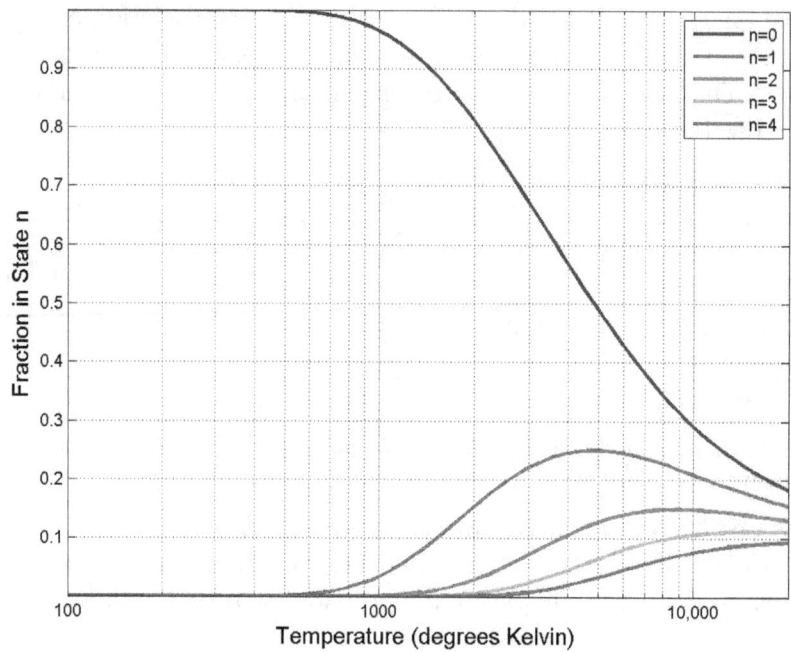

Figure 7-8: Probabilities, or fractions of the molecules, in each of the vibrational states of the diatomic nitrogen molecule. The probabilities are plotted against the temperature in degrees Kelvin.

At low temperatures when practically all molecules are in the ground state, the gas is often referred to as frozen because these vibrational modes are not a reservoir of stored energy. This is very significant for the heat capacity of gases, which will be discussed in more detail below. At room temperature, adding heat to a gas such as N_2 will increase the translational kinetic energy, but no energy will be transferred to the vibrational modes. At very high temperatures, adding heat to the gas will transfer energy to the vibrational modes.

Energy levels of rotational modes

We return to the subject of rotational modes of gas molecules. Molecules can have energy stored in rotation about their center of mass. At the atomic level, quantum mechanics applies, so the rotational energies are quantized. For linear molecules, as described above, there are always two degenerate rotational modes. Both rotational modes have the same moment of inertia. Rotational energy levels for linear molecules are described by a single quantum number, J. We will use the diatomic molecule as an example. We can visualize diatomic molecules, like N_2 in Figure 7-2, as two identical masses, m, separated by distance r. The formula for the moment of inertia is

$$I = \frac{1}{2}mr^2. \qquad (7\text{-}8)$$

The rotational energies are

$$E_J = \frac{h}{8\pi^2 I} J(J+1) \qquad (7\text{-}9)$$

where h is Plank's constant, and J is the quantum number $(0,1,2, \ldots \infty)$. The constant term in Equation (7-9) is called the *rotational constant* of the molecule. Unlike the vibrational energy levels, the ground state is 0 and the energy levels are not evenly spaced. The rotational constants for several diatomic molecules are shown in Table 7-1. Notice that the rotational constants are several orders of magnitude smaller than the vibrational energies. Also, the rotational constant for H_2 is more than an order of magnitude greater than for N_2, O_2, and NO. Observing I in Equation (7-8), the value of I for H_2 is very small since the bond length for H_2 is one of the smallest and the atomic mass is the smallest. This leads to a rather large rotational constant.

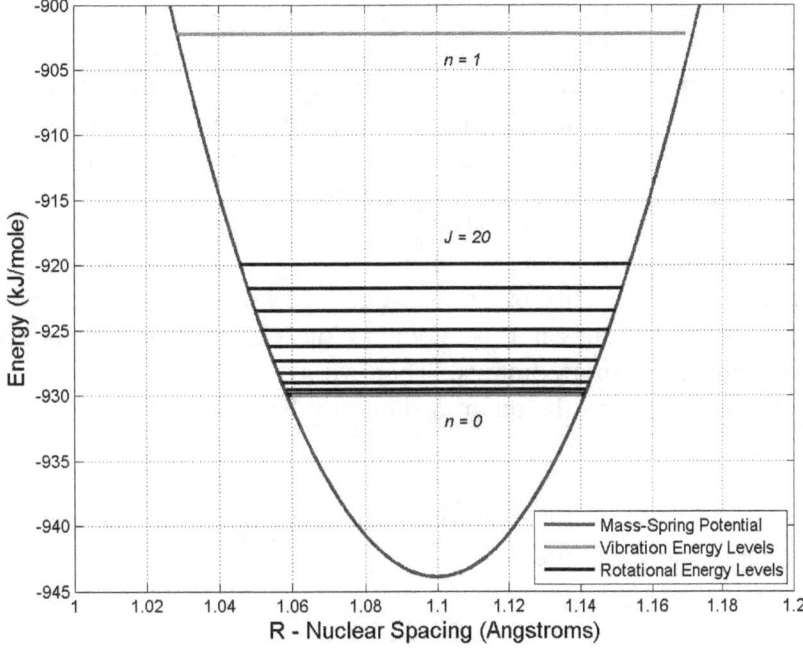

Figure 7-9: The first 20 rotational energy levels are plotted for the diatomic nitrogen molecule. Every other level is plotted for clarity. The potential energy of the mass-spring model, and two of the vibrational modes of N_2 are shown.

An illustration of the rotational energy levels for N_2 is shown in Figure 7-9 which is a zoomed plot. The rotational ground state is 0, so that state has the same energy as the vibrational ground state. The rotational energies add to the vibrational energy state, so each vibrational state has a series of rotational energies. The total energy in an internal mode for a diatomic molecule is:

$$E_{n,J} = h\nu\left(n + \frac{1}{2}\right) + \frac{h}{8\pi^2 I} J(J+1) \qquad (7\text{-}10)$$

which accounts for both quantum number associated with vibration and rotation. The rotational energy levels have much smaller jumps than the vibrational levels, so for a given vibrational energy mode, there are very many possible rotational modes, that are very close in energy. These rotational energy levels will be observed using infrared spectroscopy. A molecule may absorb or emit a photon in the infrared region of the spectrum to change vibrational mode, and it can also change rotational mode. This means that there are numerous transitions that can be measured spectroscopically. This is sometimes called *rotational-vibrational spectroscopy*, referring to the quantum change in two different internal modes.

Much like the vibrational modes, the Boltzmann distribution can predict the fraction of molecules in each of the rotational states. There are many low energy rotational modes, so at low temperatures when practically all molecules are in the ground vibrational state, many of the excited rotational states will be frozen. Going to very low temperatures, will eventually freeze out the rotational modes. For example, H_2 which has the highest rotational energy levels in Table 7-1, will have the rotational modes freeze out at temperatures below 80 °K. Other molecules, like O_2 and N_2 have freeze out temperatures in the single digits. This is significant for the heat capacity of gases. At room temperature, adding heat to a gas such as N_2 will increase the translational kinetic energy, and the rotational energy but not the vibrational energy. At extremely low temperatures, only translational modes can store added energy. The conclusion from this is the heat capacity of gases has a dependence on temperature.

Heat capacity of diatomic gases

We derive the heat capacity of diatomic gases using two methods. First, we use a classical derivation which does not use detailed knowledge of the quantized energy levels. This approach has its shortcomings. Then we use a modern derivation which applies the Boltzmann distribution to the quantized energy levels. First, we need to define heat capacity.

The *heat capacity* of a gas is the ratio of the change in heat transferred to the gas resulting in a change in temperature.

$$C = \frac{\Delta Q}{\Delta T} \quad (7\text{-}11)$$

the infinitesimal change in the heat of the gas is called ΔQ, pronounced delta-Q. If a quantity of heat, ΔQ, is transferred to the gas, the resulting temperature increase is ΔT. If heat, $-\Delta Q$, is removed from the gas, the resulting temperature decrease is $-\Delta T$. As we know from the ideal gas law a change in the internal energy of the gas can be accomplished by performing work which would be accomplished by changing the volume of the gas. For gases, one measure of heat capacity is called C_V which is the ratio of change in internal energy, ΔQ, to the resulting ΔT at *constant volume*. Since this can be viewed as a derivative, we write it as a partial derivative in T

$$C_V = \left(\frac{\partial Q}{\partial T}\right)_V. \tag{7-12}$$

Classical equipartition of energy

The concept of equipartition of energy is that all modes of a molecule (translational, vibrational and rotational) store the same quantity of energy. This is assumed to be a consequence of equilibrium. As an example, consider if one mode such as translational contained more kinetic energy than the rotational modes of a gas. Effectively, the translational modes are at a higher temperature, and individual collisions would transfer energy from the translational mode to the rotational mode until equilibrium is established.

The total internal energy of a gas, derived in Chapter 6 can be generalized to include internal modes. The total internal energy in the gas at constant volume is:

$$U_V = (\# \text{ degrees of freedom})\frac{1}{2}NkT, \tag{7-13}$$

Each mode is assumed to add the same amount to the energy of a gas. This is a consequence equipartition. For the monatomic gas which has no internal degrees of freedom, U_V is 3/2 NkT since there are 3 translational modes. This is the result which was derived in Chapter 6. For diatomic gases, the vibrational modes add NkT, which is ½ NkT for kinetic energy and ½ NkT for potential energy. The 2 degenerate rotational modes each add ½ NkT. The total for diatomic gases is then

$$U_V = \underset{\text{translation}}{\frac{3}{2}NkT} + \underset{\text{vibration}}{NkT} + \underset{\text{rotation}}{NkT} = \frac{7}{2}NkT, \tag{7-14}$$

From Equation (7-12) we can take the partial derivative of the total internal energy, keeping volume constant, to determine heat capacity.

$$C_V = \frac{\partial U_V}{\partial T} = (\# \text{ degrees of freedom})\frac{1}{2}Nk \tag{7-15}$$

The heat capacity depends upon the quantity, N, of the gas molecules, so it is usually normalized by number or molar quantity. In this case the heat capacity is

$$C_V = (\# \text{ degrees of freedom})\frac{1}{2}k \quad \text{per molecule} \tag{7-16}$$

$$C_V = (\# \text{ degrees of freedom})\frac{1}{2}R \quad \text{per mole}$$

where R is the ideal gas constant. This result is not universally correct. As was illustrated in Figure 7-8, nearly all molecules are in the vibrational ground state at room temperature. The excited states only begin to be populated above 1000 °K. So, at room temperature the heat capacity will not include the contribution due to vibration. At very low temperature, the heat capacity will not include contributions due to rotations. The molar heat capacity has different values in three different temperature regimes.

$$C_V = \frac{3}{2}R \quad \text{very low temp.} \tag{7-17}$$

$$C_V = \frac{5}{2}R \quad \text{room temp.}$$

$$C_V = \frac{7}{2}R \quad \text{high temp.}$$

Modern derivation of heat capacity

The Boltzmann distribution determines the population of the internal modes. This can be used to determine heat capacity and the temperature dependence of the heat capacity. The probability of finding a molecule in the nth vibrational state is

$$P(E_n) = \frac{e^{-E_n/kT}}{\sum_n e^{-E_n/kT}} = \frac{1}{Z} e^{-E_n/kT}. \tag{7-18}$$

where we have now defined the sum over all of the energy states to be Z. We divide each of the exponential terms by the quantity Z which is known as the *partition function*. Dividing by Z ensures that all of the probabilities sum to 1. The energy of a gas molecule is the sum over the energies in a state multiplied by the probability that the molecule is in the state. So, the total energy in the gas is:

$$U = N \sum_n E_n P(E_n) = \frac{N}{Z} \sum_n E_n e^{-E_n/kT}. \tag{7-19}$$

The heat capacity can be calculated by differentiating the total energy with respect to T. We will use H_2 as an example. We won't go into the details, but the result for C_V is plotted in Figure 7-10. The molar value of C_V is normalized by the gas constant R, so the heat capacity is plotted as a dimensionless number. This is a very interesting function. Notice that it agrees with the values 3/2, 5/2 and 7/2 in three different temperature regimes. It has the nice attribute that it smoothly transitions between these regimes. Since the values in Equation (7-17) were derived using the assumption of equipartition, this means that for practical purposes equipartition for all modes holds above 10,000 °K. Also, equipartition of energy between translational and rotational modes holds between 100-1000 °K. Notice that we can observe the freezing out behavior that was described above. The vibrational modes freeze out below 1000 °K. The rotational modes freeze out below 20 °K. Since H_2 has an unusually large rotational constant, the freezing out of rotational modes can be observed.

This seems like a very fitting way to conclude the discussion of the dynamics of gases. Deriving the heat capacity of diatomic gases is one of the triumphs of quantum mechanics. We were able to correctly calculate the heat capacity for H_2 by applying the Boltzmann distribution to the quantized vibrational energy levels. Boltzmann could not have predicted this phenomenon himself, because quantum mechanics had not yet been developed. The Boltzmann distribution and quantum mechanics form the two foundations of modern statistical mechanics. Since the Boltzmann distribution applies to quantum mechanical systems, we can certainly say he was ahead of his time.

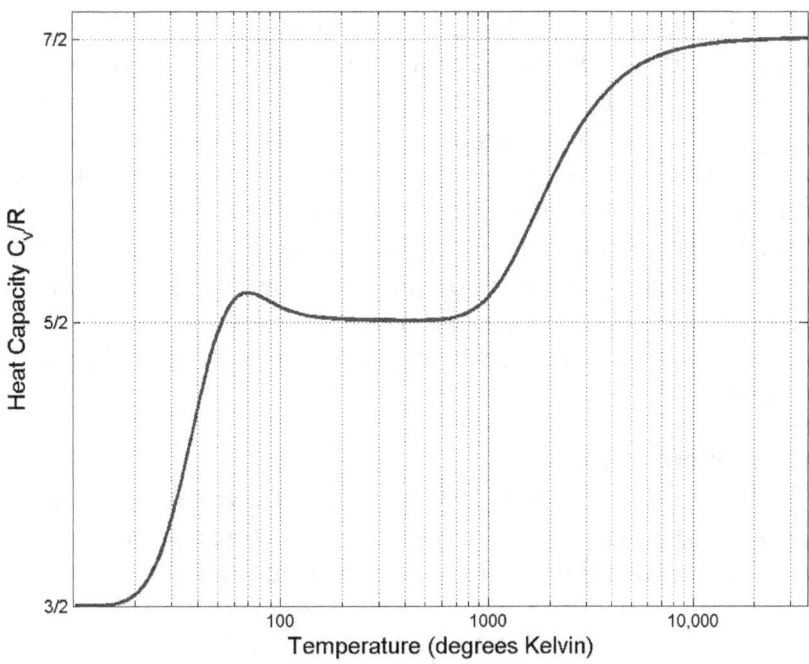

Figure 7-10: Heat capacity for hydrogen gas, H_2. The molar heat capacity is normalized by the gas constant R, so the heat capacity is a dimensionless number.

References

Halliday, D. and Resnick, R., *Fundamentals of Physics*, 2nd Ed., John Wiley & Sons, 1981.

Ma, S., *Statistical Mechanics*, World Scientific Publishing, 1985.

Rosser, W.G.V., *An Introduction to Statistical Physics*, Ellis Horwood, 1982.

WebElements: the periodic table on the web. http://www.webelements.com/

Chapter 8 : Thermodynamics and Chemical Equilibrium

Introduction

Now we go back further in time. Thermodynamics is a significant development of 19th century science. It was developed without an understanding of the atomic theory of matter, which is no indication of how useful it is. Often thermodynamics is taught by first introducing the *laws of thermodynamics*. These laws are really axioms, in the sense that they are assumed to be true, not proven true. Often students learn the laws of thermodynamics in high school or in an introductory physics class. Here we will introduce the first and second law of thermodynamic so that we can use them to derive the Gibbs energy. We will then show how the Gibbs energy allows us to determine the spontaneity of a chemical reaction. In this chapter, thermodynamics will be streamlined. We will not discuss all applications of it, but discuss the principals relevant to chemical reactions and chemical equilibrium.

The laws of thermodynamics

The *first law of thermodynamics* states that the internal energy change of a *thermodynamic system* is equal to the *heat transferred into the system* plus the *work performed on the system*. The equation for this law is written as follows

$$dU = dW + dQ = -PdV + dQ \qquad (8\text{-}1)$$

where the dU is the change in internal energy, dW the work, and dQ the heat. The small d represents an infinitesimal change in the quantity. In calculus it is called the differential. The terms P and dV will be discussed below. A thermodynamic system sounds abstract, so it is probably better to use an example system like a vessel containing an ideal gas. Such a system is shown in Figure 6-4 where an ideal gas can be compressed by a moveable piston. The piston is displaced by applying a force F which moves the piston by an infinitesimal amount dx. This displacement means that an infinitesimal amount of work is performed on the gas. The equation for the work is:

$$dW = -Fdx \qquad (8\text{-}2)$$

The pressure applied can be computed by dividing the force by the piston area, and the volume decrease can be computed by multiplying dx by the piston area as follows:

$$dW = -\frac{F}{A}Adx = -PdV \qquad (8\text{-}3)$$

which is the expected result from Equation (7-5). Note that the volume decreases ($dV<0$) when the piston is compressed, so the net work performed on the gas is positive.

Returning to Equation (7-5), we can see that if work is performed on the gas by compressing it, the change in internal energy is positive, i.e. the internal energy increases. If heat is transferred into the gas ($dQ > 0$), then the internal energy increases. From Chapter 6, it was shown the total internal energy of the ideal gas is U:

$$U = \frac{2}{3} NkT \qquad (8\text{-}4)$$

The internal energy is proportional to temperature and the number of molecules, so it should be clear that both work applied to the gas and heat transferred to the gas raise the temperature. We assume the container is insulating, and the observation time is very short compared to the amount of time required for temperature of the gas in the container to change due to conduction of heat through the insulating walls of the container. The concept that the internal energy can be increased by both heat transfer and work is referred to as *the equivalence of heat and work*.

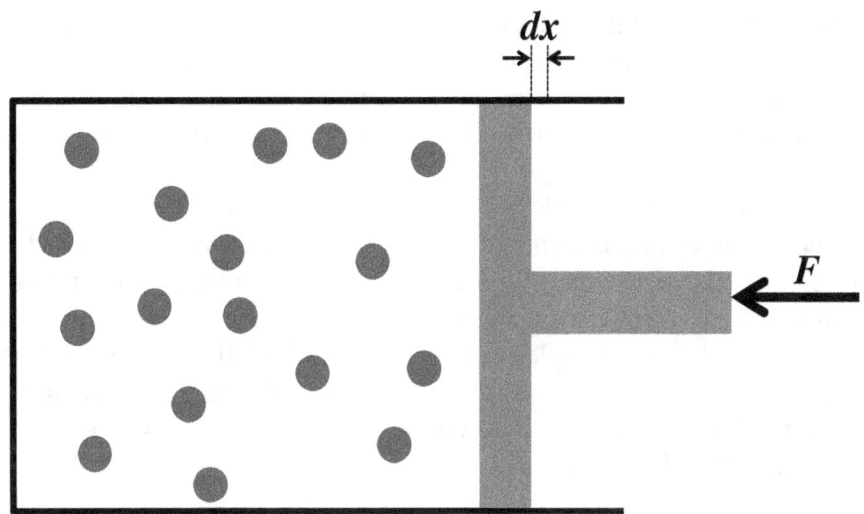

Figure 8-1: An example of a thermodynamic system. Gas molecules contained in a vessel with a movable piston, so work can be done by a force F.

Reversible and irreversible processes

A reversible process is a thermodynamic change to a system that can be performed by infinitesimal steps, which can be reversed and return the system to its original *state*. The concept of a state is important here because we need to define the state of the system and its variables in order to determine if the system has returned to the original state.

Let's use the example of an ideal gas in a vessel like Figure 6-4. The ideal gas can be compressed by moving the piston by an amount dx, which changes some of the state variables of the system. It changes reduces V, increases P, and increases T. The number of molecules N remains unchanged. Now, let's move the piston by an amount $-dx$. In that case the variables P, V and T return to their original values. The compression of an ideal gas is an example of a reversible process.

We can make the example of an ideal gas in the vessel more complicated. Let's first compress the gas by dx, which raises the temperature from T to $T + dT$. Then we will let an amount of heat dQ be transferred out of the vessel to reduce the temperature to the original T. However, the volume has not returned to its original value, so we move the piston by $-dx$. This displacement has the effect of the gas performing work, so the temperature has now been reduced to $T - dT$. By allowing the outside environment to conduct heat dQ into the gas, it will now have returned to its original, T and V. We know from the ideal gas law that this also means that the pressure, P, has returned to the original value. So, performing work or conducting heat to an ideal gas is reversible. If we perform zero work ($dW = 0$) and conduct zero heat ($dQ = 0$) to the ideal gas, it will remain in the same thermodynamic state. Such a system that has $dQ = 0$ and $dW = 0$ is defined to be a *closed system*.

By performing a cycle of processes on the ideal gas that returns it to its original state, we can see that no permanent change has occurred to the system. Chemical processes, on the other hand, are examples of permanent change as a system converts its products to reactants. These kinds of processes are considered *irreversible* in the sense that no spontaneous change will cause the reactants to revert back to products. We could perform a process which would involve work or heat input to the system that converts the products back to reactants. The system would no longer be a closed system in that case. When a closed system undergoes change to a different permanent state, the process is irreversible.

Second law of thermodynamics

The second law of thermodynamics states that *the entropy of a closed system cannot decrease*. Entropy is represented by S and infinitesimal change to entropy is dS. For a closed system any reversible cyclic processes that change a system but returns it to its original state, $dS = 0$. An example is ideal gas described above. On the other hand an irreversible process can have the entropy increase, so $dS > 0$, for irreversible processes in closed systems. So, the mathematical description of second law stated in the first sentence of this paragraph is:

$$dS \geq 0, \text{ for a closed system.} \qquad (8\text{-}5)$$

What is entropy?

Entropy has a formal definition, but sometimes it is described, somewhat informally, as a *measure of disorder*. It doesn't really make sense to use the informal description without explaining in detail how this is true, so we will dive into the formal definition. There is a classical definition of entropy and a modern definition, which relies on the quantum states of a system. We will describe both.

Classical entropy

The classical definition of entropy was first developed by German physicist Rudolph Clausius in the middle 19th century. The infinitesimal change in entropy is defined as the heat dQ transferred into the system divided by the temperature, T

$$dS = \frac{dQ}{T}. \qquad (8\text{-}6)$$

So, when a system is transformed from state A to state B, we can use calculus to integrate Equation (8-6) in order to determine the change in entropy

$$S_B - S_A = \Delta S = \int_A^B \frac{dQ}{T}. \tag{8-7}$$

Here we use the Greek delta, Δ, to represent the difference in S. This is of course pronounced, "delta S". At this point, an example we could use is the heating curve for water shown in Figure 8-2. The figure shows the temperature of one mole (18 grams) of H_2O as heat is added starting when the temperature is -25 °C. As expected the temperature rises until it reaches 0 °C where something interesting happens. Water is frozen at temperatures below 0 °C, so at this temperature adding heat melts the ice. All of the heat energy goes into melting the ice, and the temperature does not rise again until the entire 18 grams of ice is melted. This is known as the *phase transition* from solid to liquid.

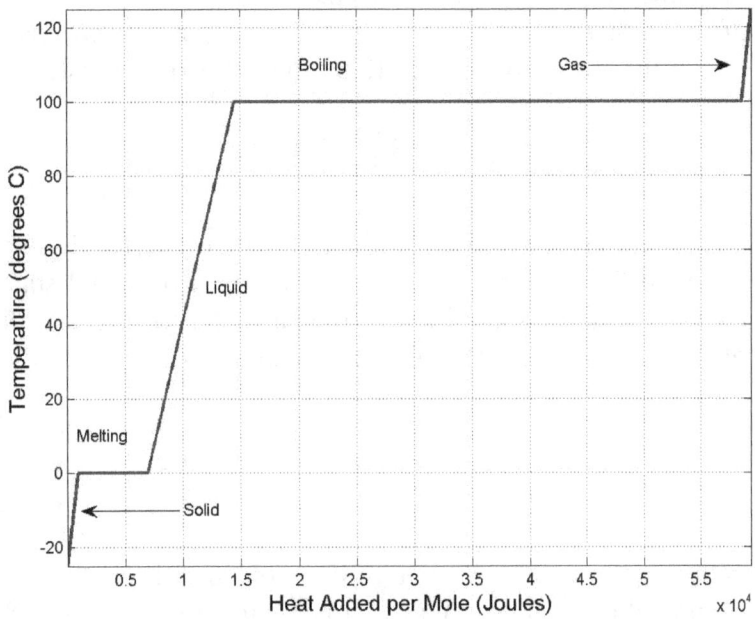

Figure 8-2: The heating curve for one mole of water. Adding heat starting with solid water (ice) will raise the temperature except at phase transitions where temperature is constant.

The temperature of 0 °C is 273.15 K, and the amount of heat to melt one mole of ice is 6010 J. Using Equation (8-7) we can see that the change in entropy is

$$\Delta S = \frac{\Delta Q}{T} = \frac{6010 \text{ J}}{273.15 \text{ K}} = 22.0 \text{ J/K} \tag{8-8}$$

We can understand why entropy is sometimes called a measure of disorder by picturing the change that takes place by melting water. The frozen water formed a crystalline solid in which the molecules were arranged in a periodic lattice. Heat transferred to the solid can increase the vibrations of the molecules in the crystal lattice, but they are still in an ordered lattice. In liquid water, the molecules can occupy many more positions and are constantly in motion changing their quantum state. For this reason the liquid state is said to be more disordered than the solid state. In order to transform a mole of water from the ordered solid state to the disordered liquid state we increase its entropy by 22.0 J/K per mole. So we can see why entropy is a measure of disorder. This will be more obvious when we introduce the definition of entropy from statistical mechanics.

Boltzmann definition of entropy

The Austrian physicist Ludwig Boltzmann was introduced in Chapter 6 as one of the most important developers of kinetic theory. He developed a definition of entropy based upon statistical mechanics typically referred to as the Boltzmann definition of entropy

$$S = k \log \Omega \tag{8-9}$$

where k is the Boltzmann constant and Ω is the number of states of the system. This equation is on Boltzmann's tombstone (it uses the variable W rather than Ω). The modern interpretation of the number of states is the number of quantum mechanical states. It is possible to derive from the expression in (8-9) the following equation which is a sum over all states n:

$$S = -k \sum_n p_n \log p_n \tag{8-10}$$

where the variable p_n is the probability of finding the thermodynamic system in a state n. We'll use the example from the previous chapter of the vibrational states in a diatomic nitrogen gas. These probabilities can be determined by the Boltzmann distribution. Recall that the probability of a molecule with vibrational energy E_n is

$$p_n = \frac{e^{-E_n/kT}}{\sum_n e^{-E_n/kT}}. \tag{8-11}$$

The probability represents the fraction of the molecules in each of the energy states, so the sum of all probabilities must equal 1. It is assumed that the probability of finding each N_2 molecule in a given vibrational states is independent of the other molecules. The entropy for a number N of nitrogen gas molecules is then

$$S = -kN \sum_n p_n \log p_n \tag{8-12}$$

where the probabilities are from Equation (7-7). Note that this example is only the contribution to S from the vibrational states. Obviously, we would have to add in the translational and rotational modes in order to calculate the total entropy, but this example is intended to illustrate the concept without being too complicated.

A plot of the probabilities for the N_2 molecule is shown in Figure 8-3 (top). The entropy from Equation (8-10) is shown in the bottom plot. At room temperature and below most molecules are in the ground vibrational state which means the probability of finding the system in the ground state is 1 or very close to 1. The system in the ground state gives entropy value of 0 because in Equation (8-10), the value of log(1) is 0. So, the sum over all states is 0. As the temperature increases, a fraction of the molecules are raised into excited states. Then the sum in Equation (8-10) has a non-zero value. As the temperature approaches even higher temperatures, the fraction of molecules in each of the states will approach the same value and more and more states have a non-zero probability of being populated. The system of N_2 molecules is most ordered at low temperature when all of the molecules are in the ground vibrational state. Disorder increases with temperature as the probability of finding a molecule in one of many excited states becomes larger.

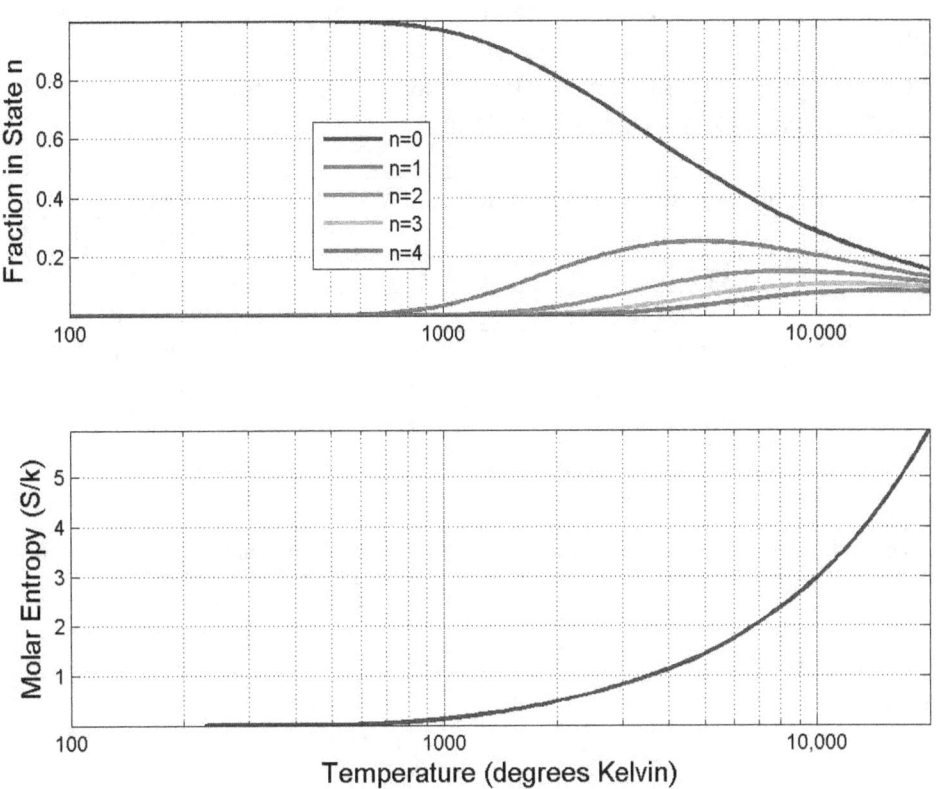

Figure 8-3: (Top) Probabilities, or fractions of the molecules, in each of the vibrational states of the diatomic nitrogen molecule. The probabilities are plotted against the log of temperature in degrees Kelvin. (Bottom) The molar entropy for the vibrational states which increases with temperature and goes to zero when the vibrational modes freeze out.

At low temperatures when all molecules are in the same ground state Equation (8-12) requires that the entropy is zero; however, this is a more general principle. At temperatures of absolute zero, any system will be in its ground state, so the number of states will be 1. From the Boltzmann entropy definition $\log\Omega$ will be zero, so $S = 0$ at temperature equal to absolute zero. This statement ($S = 0$ at $T = 0$) is known as the *third law of thermodynamics*.

State variables

The Boltzmann definition of entropy calculates entropy from the bottom up so to speak. First, the statistical mechanics of the system are defined, and from that the entropy can be calculated. The classical definition on the other hand uses measureable variables T and Q to calculate the entropy. Entropy is known as a *state variable* because it characterizes the current state of a thermodynamic system independent of how the path taken to that state. Entropy seems somewhat abstract compared to other intuitive state variables like volume and temperature, so it may not be obvious that it is also an *extensive property* which means the value depends upon the size of the system. The opposite type is an *intensive property* which does not depend upon the size of the system.

Table 8-1 lists some of the state variables that are important for basic chemistry. It should be obvious that quantities like volume and number of particles are extensive, and equally obvious that temperature and pressure are intensive. Heat capacity and entropy are extensive properties that are often normalized by the number of moles. We typically refer to molar entropy, which is per mole of the substance.

State Variable Name	Variable Symbol	Type
Entropy	S	Extensive/Intensive
Pressure	P	Intensive
Temperature	T	Intensive
Volume	V	Extensive
Particle Number	N	Extensive
Internal Energy	U	Extensive
Heat Capacity	C	Extensive/Intensive

Table 8-1: List of common state variables used in this text. Note some extensive properties are often represented as molar quantities, and molar quantities are necessarily intensive properties.

Second law, again

Now, we define the second law again, but for systems including those in which heat can be transferred:

$$dS \geq \frac{dQ}{T}. \tag{8-13}$$

The equality holds for reversible processes, and the inequality holds for irreversible processes. Spontaneous chemical reactions are the irreversible processes we are interested in, and this will be discussed next.

Gibbs energy and spontaneity

The Gibbs energy is named after the American physicist J. Willard Gibbs. It is sometimes referred to as the Gibbs free energy which is somewhat anachronistic. Gibbs energy, represented by G, is defined:

$$G = U + PV - TS \tag{8-14}$$

where U is internal energy, P is pressure, V is volume, T is temperature, and S is entropy. Since G is a function of state variables, it is also a state variable. Also, since it is proportional to extensive properties, G is an extensive property.

Reactions that go forward are irreversible, so for reactions the second law of thermodynamics tells us

$$dS > \frac{dQ}{T}. \tag{8-15}$$

Furthermore, we know from re-arranging the first law that

$$dQ = dU + PdV, \tag{8-16}$$

and substituting Equation (8-16) into (8-15) gives us

$$dU + PdV - TdS < 0, \tag{8-17}$$

Since we are interested in chemical reactions at constant temperature and pressure, the change in Gibbs energy is

$$dG = dU + PdV - TdS, \tag{8-18}$$

and the criterion for spontaneous reactions is

$$dG < 0 \tag{8-19}$$

We have just derived the condition for spontaneity. For any given reaction we replace the infinitesimal dG with a finite ΔG representing the change in Gibbs energy of molar quantities involved in the reaction. For constant temperature and pressure conditions

$$\Delta G = \Delta U + P\Delta V - T\Delta S. \tag{8-20}$$

Note that from the first law, $\Delta U + P\Delta V = \Delta Q$, so the heat released (or absorbed) during the reaction is from the change in internal energy plus work done. The heat released during a reaction is correctly called the *heat of reaction* or *enthalpy of reaction*, and this is the quantity we can measure and calculate most easily. We will use the term enthalpy, H, which is defined as

$$H = U + PV, \tag{8-21}$$

so the criterion for spontaneity of reactions is succinctly written

$$\boxed{\Delta_r G = \Delta_r H - T\Delta_r S < 0.} \tag{8-22}$$

The change in G is now represented as Δ_r where the subscript r refers to the reaction, so $\Delta_r G$ is known as the *Gibbs energy of reaction*. The chemical definition of *spontaneous reaction* is a reaction where the change in Gibbs energy ($\Delta_r G$) is negative for the reaction. The plain language definition is a reaction that proceeds forward without any outside intervention. The rate of reaction is independent of spontaneity, so spontaneous reactions can be very slow, very fast or any rate in between.

Typically, analysis takes place at *standard ambient temperature and pressure*, referred to as STAP. Standard ambient temperature is 298.15 K and standard ambient pressure is 1 bar. Under STAP conditions, the enthalpy of reaction is represented as $\Delta_r H^o$ and is referred to as the *standard enthalpy of reaction*. It has units of Joules/mole, or kiloJoules/mole, so $\Delta_r H^o$ is an intensive property because it has been normalized by molar quantities. The entropy change is denoted as $\Delta_r S^o$ and is called the *standard entropy of reaction*. The value of $\Delta_r S^o$ is measured under standard conditions, which means pure compounds and is computed by subtracting the entropy of the reactants from the entropy of the products. From these quantities we can calculate the *standard Gibbs energy of reaction*, $\Delta_r G^o$, as discussed below.

Spontaneous exothermic and endothermic reactions

The criterion for spontaneity (Equation (8-22)) is a significant development in the history of chemistry. This was discovered by the German physical chemist Walter Nernst and published in 1906. This contribution and his later work led to the Nobel prize in chemistry awarded in 1920.

Chemical reactions that release energy (typically in the form of heat) during the reaction are called *exothermic reactions*, and reactions that absorb heat from their surroundings are *endothermic reactions*. At the time that Nernst discovered what was called his heat theory, it was believed by many that a reaction was spontaneous if it progressed from a higher energy state to a lower energy state, releasing heat in the process. It was therefore considered a puzzle prior to Nernst's discovery that spontaneous endothermic reactions existed.

Example reactions

The following reaction is solid (s) ammonium nitrate dissolving in water to form an aqueous (aq) solution. In the process energy is absorbed from the system's surrounding environment, so it is endothermic. This reaction takes place in the instant cold packs that are sold for first aid or sports injuries. The heat absorbed as the reaction progresses reduces the temperature of the water in the pack, cooling it significantly.

$$NH_4NO_3 \text{ (s)} \rightarrow NH_4^+ \text{ (aq)} + NO_3^- \text{ (aq)}$$

The standard enthalpy of reaction ($\Delta_r H^o$) is positive, consistent with an endothermic reaction. The value of 25.7 kJ/mole is the standard enthalpy of reaction. The following calculation for Gibbs energy is:

$\Delta_r H^o = +25.4$ kJ/mole

$\Delta_r S° = +0.109$ kJ/mole-Kelvin
$T = 298.15$ K

$$\Delta_r G° = \Delta_r H° - T\Delta_r S° = -4.5 \text{ kJ/mole}$$

So the Gibbs energy is negative indicating the reaction is spontaneous. This confirms that a reaction can certainly be endothermic and spontaneous. This can be the case as long as the second term in the Gibbs energy ($-T\Delta_r S°$) is negative and greater in magnitude than the change in enthalpy. For reactions that are dissolution of solids into liquid, the $\Delta_r H$ is more specifically called *enthalpy of solution*.

Another example of a solid dissolving in solution is the instant hot pack. These are similar to the instant cold packs, only a solid which is exothermic when dissolved in water is used. One common example is the salt calcium chloride dissolving in water.

$$CaCl_2 \text{ (s)} \rightarrow Ca^{2+} \text{ (aq)} + 2Cl^- \text{ (aq)}$$

$\Delta_r H° = -81.4$ kJ/mole
$\Delta_r S° = 0.0615$ kJ/mole-Kelvin
$T = 298.15$ K

$$\Delta_r G° = \Delta_r H° - T\Delta_r S° = -99.7 \text{ kJ/mole}$$

Note that because the reaction is exothermic and the change of entropy is positive, the Gibbs energy is negative for any temperature (as long as the solvent water is a liquid). Remember that temperatures on the Kelvin scale are always non-negative. The reaction will be spontaneous at all temperatures.

Another example is oxidation of iron leading to rust. Since the reactants on the left side include gas molecules, we expect that the reaction leads to reduction in disorder. The negative change in entropy confirms that assumption.

$$4Fe \text{ (s)} + 3O_2 \text{ (g)} \rightarrow 2Fe_2O_3 \text{ (s)}$$

$\Delta_r H° = -824$ kJ/mole
$\Delta_r S° = -0.549$ kJ/mole-Kelvin
$T = 298.15$ K

$$\Delta_r G° = \Delta_r H° - T\Delta_r S° = -660 \text{ kJ/mole}$$

The reaction is spontaneous because $\Delta_r G°$ is negative. Even though this reaction has the largest Gibbs energy change of all of the examples, it is a very slow reaction. Rust forms very slowly. This confirms the earlier observation that reaction rate is independent of spontaneity. Another observation is that $\Delta_r S°$ has one of the larger negative values; however, the very large enthalpy change of reaction is the dominant term.

Effect of temperature on spontaneity

The formula for Gibbs energy is a linear function of temperature, so spontaneity of a reaction can depend upon the temperature of the system. In fact we can control the spontaneity of a reaction by controlling the temperature. As an example, we will analyze an endothermic reaction which will have a positive $\Delta_r H°$. Furthermore, we'll assume the reaction also has positive $\Delta_r S°$. The change in entropy and enthalpy change very little over temperature, so a reasonable approximation is to keep those variables constant with temperature. In this case, the change in Gibbs energy will look like Figure 8-4. In the figure we have used the decomposition reaction of sodium bicarbonate

$$4NaHCO_3 \, (s) \rightarrow Na_2CO_3 \, (s) + CO_2 \, (g) + H_2O \, (g)$$

This reaction has $\Delta_r H° = 136$ kJ/mole, and $\Delta_r S° = +0.33$ kJ/mole-Kelvin, so temperature dependent change in Gibbs energy is

$$\Delta_r G = \Delta_r H° - T\Delta_r S° = 136 - 0.33T \tag{8-23}$$

which crosses the horizontal axis ($\Delta_r G = 0$) at a temperature T^*

$$T^* = \frac{\Delta_r H°}{\Delta_r S°}. \tag{8-24}$$

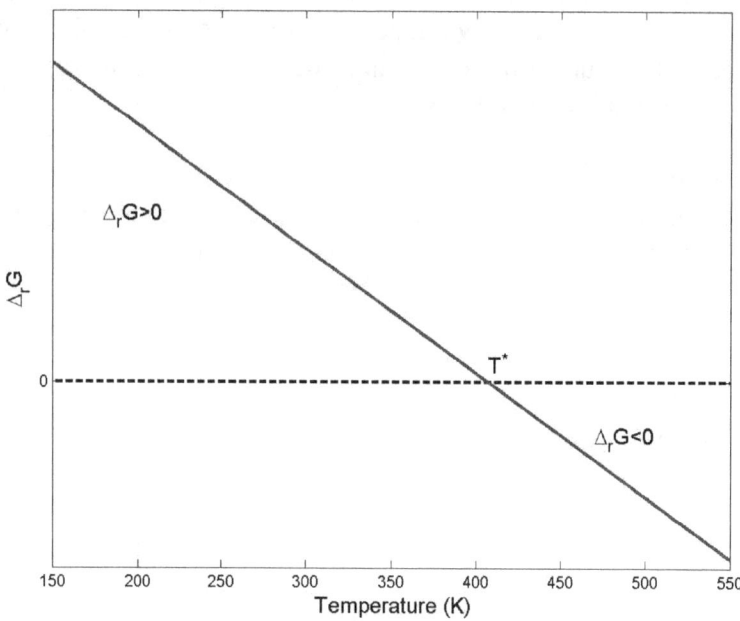

Figure 8-4: The temperature dependence of the Gibbs energy change of reaction ($\Delta_r G$) for decomposition of sodium bicarbonate. At low temperatures the reaction is not spontaneous, but above temperature $T^* = 407.2$ K the reaction is spontaneous.

As can be seen in Figure 8-4, for temperatures above the value of $T^* = 407K$ the reaction is spontaneous because $\Delta_r G$ is negative, and $\Delta_r G$ is positive at lower temperatures.

This means we can control the reaction. By raising the temperature the reaction goes forward, at room temperature (298K) sodium bicarbonate is stable and does not decompose.

Why do we care about the decomposition of sodium bicarbonate? It is more commonly known as baking soda and is used in cookie and cake batter. The temperature of 407K corresponds to 273 °F, so at typical oven baking temperatures of 350-450 °F, sodium bicarbonate will decompose and release significant quantities of CO_2. The CO_2 gas will form bubbles in the baking batter making the cake lighter and fluffy. It is interesting that the question of why cakes rise during baking is related to basic thermodynamics like entropy, enthalpy and spontaneity.

Calculating standard entropy of reaction

The standard entropy of reaction is always for molar quantities. It is calculated by using the *standard molar entropy*, $S°$. This value is an absolute measure of the entropy contained in one mole of the compound under standard conditions. A pure element under standard conditions has a non-zero value of $S°$. The standard entropy of reaction, $\Delta_r S°$, is calculated using the following formula:

$$\Delta_r S° = \sum_{\text{Products}} S° - \sum_{\text{Reactants}} S° \qquad (8\text{-}25)$$

Many tables containing standard molar entropy for various compounds have been compiled. For the example of solid (s) ammonium nitrate dissolving in water to form an aqueous (aq) solution:

$$NH_4NO_3 \text{ (s)} \rightarrow NH_4^+ \text{ (aq)} + NO_3^- \text{ (aq)}$$

$S°$ for one mole of solid NH_4NO_3 is 151.1 J/K (per mole), $S°$ for one mole of an aqueous solution of NH_4^+ is 113.4 J/K, and $S°$ for one mole of an aqueous solution of NO_3^- is 146.4 J/K, so

$$\Delta_r S° = +151.1 \text{ J/K} - 113.4 \text{ J/K} - 146.4 \text{ J/K} = -108.7 \text{ J/K}$$

For the example of a solid calcium chloride dissolving in water,

$$CaCl_2 \text{ (s)} \rightarrow Ca^{2+} \text{ (aq)} + 2Cl^- \text{ (aq)}$$

$S°$ for one mole of solid $CaCl_2$ is 104.6 J/K $S°$ for one mole of an aqueous solution of Ca^{2+} is 53.1 J/K, and $S°$ for one mole of an aqueous solution of Cl^- is 56.5 J/K, so

$$\Delta_r S° = +104.6 \text{ J/K} - 53.1 \text{ J/K} - 2(56.5) \text{ J/K} = -61.5 \text{ J/K}$$

For the example of a solid iron oxidizing to form rust,

$$4Fe \text{ (s)} + 3O_2 \text{ (g)} \rightarrow 2Fe_2O_3 \text{ (s)}$$

$S°$ for one mole of solid Fe is 27.3 J/K, $S°$ for one mole of gaseous O_2 is 205.0 J/K, and $S°$ for one mole of solid Fe_2O_3 is 87.4 J/K, so

$$\Delta_r S° = 2(87.4) \text{ J/K} - 4(27.3) \text{ J/K} - 3(205.0) \text{ J/K} - = -549.4 \text{ J/K}$$

Calculations of this type are very straightforward as these examples indicate.

Calculating standard enthalpy of reaction

There are two main methods for calculating the standard enthalpy of reaction, $\Delta_r H°$, which we'll discuss here. The methods are the *bond enthalpy* and the *heat of formation*. There are tables listing enthalpy for chemical compounds but not typically for reactions. It is convenient to know multiple methods to calculate $\Delta_r H°$ depending upon the information available.

First method: bond enthalpy

Bond enthalpy is the heat released when a bond is dissociated. As described in Equation (8-21) above, the change in energy (or heat) is equal to the change in enthalpy minus the work done:

$$\Delta U = \Delta H - \Delta(PV). \qquad (8\text{-}26)$$

The bond enthalpy ignores the work performed, $\Delta(PV)$, which differs at most in gases by 1-2 percent. It is easy to measure change in heat when bonds are broken, so the standard bond enthalpy $\Delta H°$ is measured for molar quantities under standard conditions (1bar pressure, 298.15 K temperature).

As described above an exothermic reaction has $\Delta H°$ positive so energy is released when the bond is broken. An endothermic reaction requires an input of heat to break the bond. From this we conclude that when the bonds are stronger in the reactants, the reaction is endothermic. When the bonds are stronger in the products, the reaction is exothermic. There are tables of standard bond enthalpies, so all types of bonds can be looked up. For a particular bond, *e.g.* C–H bond, the bond enthalpy varies for different molecules. Bond enthalpies are typically within 8% of average, so tables average the bond enthalpy over all of the values for different molecules. The calculation for $\Delta_r H°$ subtracts the sum of all of the bond enthalpies in the reactants from that of the products:

$$\Delta_r H° = \sum_{\text{Bonds in Products}} \Delta H° - \sum_{\text{Bonds in Reactants}} \Delta H°. \qquad (8\text{-}27)$$

For example, we can calculate $\Delta_r H°$ for the oxidation of glucose. The structural description of glucose is shown below in Figure 8-5. The oxidation of glucose is one of the most important biochemical reactions:

$$C_6H_{12}O_6 + 6O_2 \rightarrow 6CO_2 + 6H_2O \ .$$

From the structure and the chemical equation above, we can sum the bond enthalpy on the product and reactant side. The bonds of interest in the reactant side are 5 O–H bonds, 5 C–O bonds, 7 C–H bonds, 5 C–C bonds and 1 C=O bond from the glucose molecule. There are 6 O=O bonds from the gaseous oxygen. All of these bonds are broken. The product side has 12 O–H plus 12 C–O bonds which are formed.

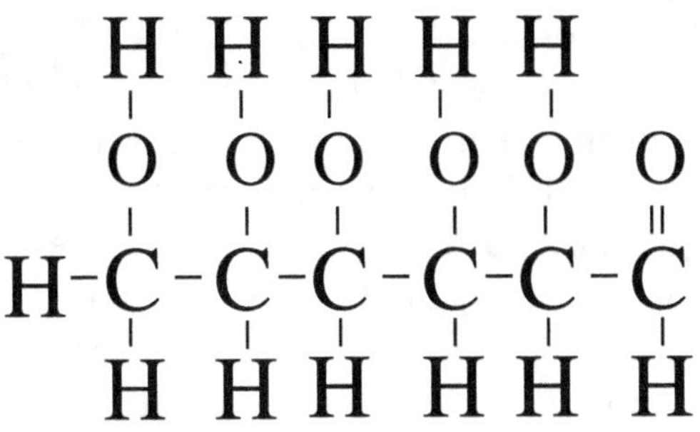

$$
\begin{array}{cccccc}
H & H & H & H & H & \\
| & | & | & | & | & \\
O & O & O & O & O & O \\
| & | & | & | & | & \| \\
H-C-&C-&C-&C-&C-&C \\
| & | & | & | & | & | \\
H & H & H & H & H & H
\end{array}
$$

Figure 8-5: The structure of glucose, (C$_6$H$_{12}$O$_6$).

Using average bond enthalpies from standard tables, the sum of products is:

$$\sum_{\text{Bonds in Products}} \Delta H^\circ = 12{,}450 \text{ kJ/mole} \tag{8-28}$$

and the sum of the reactants are:

$$\sum_{\text{Bonds in Reactants}} \Delta H^\circ = 15{,}190 \text{ kJ/mole} \,. \tag{8-29}$$

The calculated enthalpy of reaction is then −2740 kJ/mole which is within 3% of the experimentally determined value of $\Delta_r H^\circ$ which is −2816 kJ/mole.

Second method: heat of formation

The molar heat of formation is the amount of heat released (or absorbed) when one mole of a compound is formed from pure elements in the elements' standard state. The standard state refers to standard ambient temperature and pressure (STAP) as well as the form that stable standard state of the element under STAP conditions. For example, water is formed from H$_2$ and O$_2$, so diatomic gas is the standard state of the two reactants. It should be obvious that the heat of formation of elements in the standard state is 0 kJ/mole. There are standard tables of heat of formation as well since it is readily measured experimentally. Standard enthalpy of formation, ΔH_f°, is considered equal to the heat of formation. As for the bond enthalpy, the work performed, $\Delta(PV)$, is ignored.

For the example of glucose oxidation, the heat of formation for oxygen is 0, because it is already in its most stable state. So the sum of heat of formation of products minus that of reactants is:

$$\Delta_r H^\circ = \Delta H_f^\circ(6CO_2) + \Delta H_f^\circ(6H_2O) - \Delta H_f^\circ(C_6H_{12}O_6) - \Delta H_f^\circ(6O_2)$$

$$\Delta_r H^\circ = 6(-393.51) + 6(-285.83) - (-1271) - 6(0) = -2805 \text{ kJ/mole}$$

This calculation has excellent agreement with experimental determined value of $\Delta_r H^\circ$ which is -2816 kJ/mole.

Nature of chemical equilibrium

Walter Nernst is known for developing the criterion for spontaneity, but he was also involved in the theory of *chemical equilibrium*. Nernst consulted to industry and became wealthy from patenting chemical inventions, so he was very much aware of the importance of the theory of equilibrium both scientifically and economically. For example, an important industrial process, the *Haber process*, produces ammonia from nitrogen and hydrogen gas. Nitrogen gas is nearly inert under normal conditions, since the triple bond for diatomic nitrogen is especially strong. The development of an energy efficient method for producing ammonia was commercially significant since ammonia can be used to manufacture fertilizers and explosives.

The Haber process combines gaseous hydrogen and nitrogen at high temperatures (around 500 C°) in the presence of a catalyst, usually iron, to produce ammonia (NH_3).

$$N_2 + 3H_2 \rightarrow 2NH_3$$

In the chemical reactor built for this process, some amount of ammonia will dissociate back into nitrogen and hydrogen.

$$2NH_3 \rightarrow N_2 + 3H_2$$

Thus far in the book we have only discussed forward reactions, but what is described above is the reverse reaction. Forward and reverse reactions are proceeding at the same time. When the amount of ammonia produced is equal to the amount dissociated, we say the equation is in chemical equilibrium. We write the combined forward and reverse reactions as follows.

$$N_2 + 3H_2 \rightleftharpoons 2NH_3$$

In general a chemical reaction will typically proceed for some amount of time and then appear to stop before the reactants are used up. When the reaction stops it has reached equilibrium. *Equilibrium is when the concentration of the products and reactants are unchanging over the observation time.* For chemical reactions that are reversible, this means that the rate of the forward reaction is equal to the rate of the reverse reaction, which is a consequence of the principal of detailed balance. The economic significance should be clear. It is desirable to know if the forward reaction will proceed and what percentage of the reactants are converted to the product. The *equilibrium equation* describes the conversion of reactants to products.

Equilibrium equation and rates of reaction

It may be best to derive the equilibrium equation for a specific reaction and then the general equation will be more easily understood. An interesting reaction that occurs in the gas phase is the conversion reaction between the gas nitrogen dioxide (NO_2) and the gas dinitrogen tetroxide (N_2O_4). A mixture of the gases NO_2 and N_2O_4 at low temperatures is nearly colorless, but at high temperatures it acquires a deep red color. The color change is because the two gases are in equilibrium according to the equation:

$$N_2O_4 \rightleftharpoons 2NO_2$$
$$\text{colorless} \quad \text{red}$$

The concentrations of the two gases are represented as follows:
$[N_2O_4]$ = concentration of dinitrogen tetroxide in moles per liter
$[NO_2]$ = concentration of nitrogen dioxide in moles per liter

Let's determine the rates of the reactions. The forward reaction, the decomposition of N_2O_4, occurs when the bond between the two N atoms is broken. Only a fraction of the molecules in an observation time have sufficient energy to break this bond. We'll call the fraction of all N_2O_4 molecules that decompose in one second k_F. The subscript F is used because this is the forward reaction. The fraction of the molecules multiplied by the concentration of the molecules in moles/liter gives the number of moles decomposed per second. We call this quantity the rate of the reaction:

$$k_F[N_2O_4] = \text{moles of } N_2O_4 \text{ decomposed per second.} \qquad (8\text{-}30)$$

Note that the rate depends upon concentration of the reactant. Let's determine the rate of the reverse reaction. The reverse reaction occurs when two NO_2 molecules collide and they combine to form one N_2O_4 molecule. The probability that one particular NO_2 molecule collides with all other NO_2 molecules depends upon the concentration of NO_2. For example, if we double the concentration of NO_2 molecules, the probability of that particular molecule colliding with another NO_2 molecule is increased by 2. The total number of collision is proportional to $[NO_2]^2$. The fraction of all the collisions that lead to a formation of one N_2O_4 molecule is k_R, where the subscript R stands for the reverse reaction. The rate of the reverse reaction in moles formed per second is:

$$k_F[NO_2] = \text{moles of } NO_2 \text{ decomposed per second.} \qquad (8\text{-}31)$$

The probability of a collision is proportional to $[NO_2]^2$. It is not a coincidence that the exponent is 2 and the reverse reaction combines 2 quantities of NO_2; it is because two NO_2 molecules collide in the reaction. If 3 molecules were required to react, the exponent will be 3. This will be generalized below.

The coefficients k_F and k_R are called the reaction rate constants for two opposing reactions. They are constant at constant temperature, but typically increase with increasing temperature. Once the reaction has reached equilibrium, the forward reaction rate is equal to the reverse reaction rate. This is a restatement of the definition of equilibrium:

$$k_R[NO_2]^2 = k_F[N_2O_4]. \qquad (8\text{-}32)$$

It leads to the derivation of the *equilibrium equation*:

$$K = \frac{k_F}{k_R} = \frac{[NO_2]^2}{[N_2O_4]}. \qquad (8\text{-}33)$$

where K is called the *equilibrium constant*. The equilibrium constant is independent of the pressure or concentration of the reacting substances. Note that the concentration of products at equilibrium is in the numerator, and reactant concentration is the denominator. When K > 1, we can expect a high concentration of products. When K < 1, we can expect a low concentration of products.

General equation for equilibrium
Let's take a generic chemical reaction

$$\alpha A + \beta B + \ldots \rightleftharpoons \chi C + \delta D + \ldots \qquad (8\text{-}34)$$

where A, B, C and D represent some chemical substances and the Greek letters are the quantities. Then the equilibrium constant is:

$$K = \frac{[C]^\chi [D]^\delta \cdots}{[A]^\alpha [B]^\beta \cdots}. \qquad (8\text{-}35)$$

Equation (8-35) is correct if the reaction takes place in the gas phase or the liquid phase. In either case the concentration of [A] is moles/volume. Generally, *when K > 1, we can expect a high concentration of products. When K < 1, we can expect a low concentration of products.*

Chemical equilibrium and Gibbs energy

In order for the forward reaction to progress spontaneously, the Gibbs energy has to be negative, $\Delta_r G < 0$, as discussed in the first half of this chapter. Likewise, the reverse reaction will progress spontaneously when $\Delta_r G > 0$. This should be intuitive since if the equation is turned around, the reactants become the products and the sign of $\Delta_r G$ changes. *When a reaction is not in equilibrium* the Gibbs energy for the reaction depends upon the concentration of the reactants and products:

$$\Delta_r G = \Delta_r G^\circ + RT \ln(Q) \qquad (8\text{-}36)$$

where R is the gas constant, T is temperature, and

$$Q = \frac{[C]^\chi [D]^\delta \cdots}{[A]^\alpha [B]^\beta \cdots} \qquad (8\text{-}37)$$

which has the same mathematical expression as the equilibrium constant above. The difference is that the reaction has not progressed to equilibrium in this case.

Why is $\Delta_r G$ variable?

The temperature of the system and the relative concentrations of reactants and products determines the Gibbs energy, $\Delta_r G$. As an example, let's use the decomposition of the gas dinitrogen tetroxide (N_2O_4) which we discussed above. The value of $\Delta_r G$ is computed using Equation (8-36) for this reaction

$$\Delta_r G = \Delta_r G^\circ + RT \ln\left(\frac{[NO_2]^2}{[N_2O_4]}\right). \tag{8-38}$$

The results are plotted in Figure 8-6. The Gibbs energy of reaction, $\Delta_r G$, is plotted as a function of the concentration of N_2O_4. What isn't shown is the concentration of NO_2, but that concentration rises in proportion to the dissociation of N_2O_4. At low concentrations of the reactant N_2O_4, the concentration of the product NO_2 is high, so the reverse reaction progresses and $\Delta_r G$ is positive. For high concentrations of N_2O_4 the value of $\Delta_r G$ is negative. In this case, since N_2O_2 concentration is very high, the forward reaction proceeds and more reactant dissociates to form more product, NO_2.

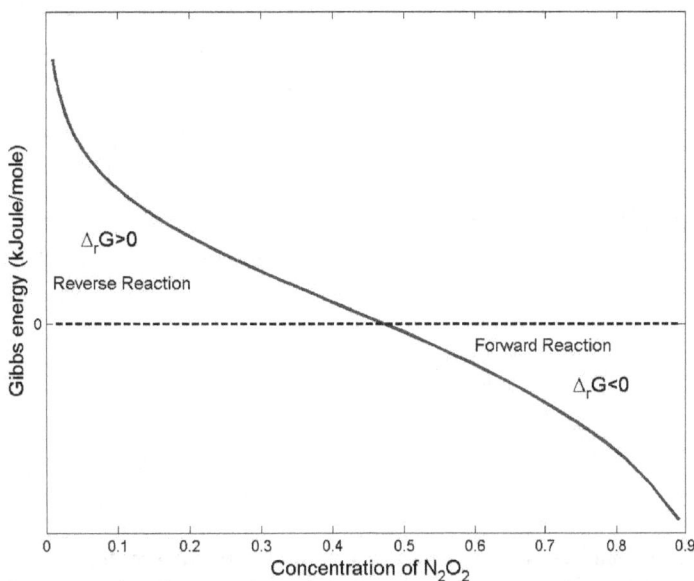

Figure 8-6: The dependence of the Gibbs energy change of reaction ($\Delta_r G$) for decomposition of N_2O_4. At low concentrations of the reactant N_2O_4 the concentration of the product NO_2 is high, so the reverse reaction progresses and $\Delta_r G$ is positive. For higher concentrations of N_2O_4 the value of $\Delta_r G$ is negative so the forward reaction progresses.

At a particular concentration the reaction will be in equilibrium which is when $\Delta_r G = 0$. This is when the forward and reverse reaction rates are exactly balanced. If we take Equation (8-36), and set $\Delta_r G = 0$, we get the following equation:

$$\Delta_r G^\circ = -RT \ln(Q). \tag{8-39}$$

At equilibrium $Q = K$ where K is the equilibrium constant in Equation (8-35). This leads to an expression relating the equilibrium constant to the standard Gibbs energy of reaction:

$$K = \exp\left(-\frac{\Delta_r G^\circ}{RT}\right). \tag{8-40}$$

As discussed above, values of $K < 1$ indicate low concentration of products, but when $K > 1$ there are high concentration of products. Returning to the example of dissociation of N_2O_4, Equation (8-40) is plotted for that reaction. The equilibrium constant, K is shown plotted against temperature. It is clear that the value of K is near 0.1 at room temperature. Thus a glass vessel of N_2O_4 should appear colorless at room temperature since very little will have been dissociated to form NO_2, which has a red color. As the vessel is heated to high temperatures, the equilibrium state has higher concentrations of NO_2, and the gas does indeed acquires a reddish color as temperature increases.

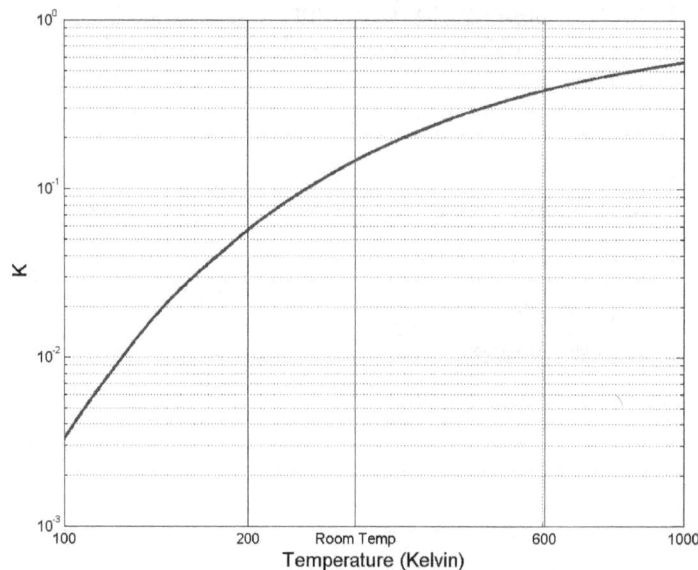

Figure 8-7: The dependence of the equilibrium constant K on temperature for the dissociation of N_2O_4. At room temperature the value of K is near 0.1 which indicates very little conversion of the product to the reactant NO_2. At high temperature more product is formed causing the gas to acquire a reddish color.

As stated above, $Q = K$ at equilibrium. When $Q < K$, the reaction will shift toward products and when $Q > K$, the reaction will shift toward reactants.

Temperature dependence of chemical equilibrium

Equation (8-39) illustrates the relationship between Gibbs energy and the equilibrium constant. Equation (8-22) illustrates the relationship between Gibbs energy and the other thermodynamic properties entropy, enthalpy and temperature. Combining these equations leads to:

$$\Delta_r G^\circ = -RT \ln(K) = \Delta_r H^\circ - T\Delta_r S^\circ. \tag{8-41}$$

This equation leads to the following equation for the equilibrium constant, K:

$$\ln(K) = -\frac{\Delta_r H^\circ}{RT} + \frac{\Delta_r S^\circ}{R}, \qquad (8\text{-}42)$$

so K changes with a change in T. The standard enthalpy of reaction and the standard entropy of reaction are independent of temperature, so the $1/T$ in the first term of the equation above is the only term that changes with temperature. The equilibrium constant at two different temperatures can be compared using Equation (8-42). The following equation shows the natural log of the ratio of two equilibrium constants at temperatures T_1 and T_2:

$$\ln\left(\frac{K_2}{K_1}\right) = -\frac{\Delta_r H^\circ}{R}\left(\frac{1}{T_2} - \frac{1}{T_1}\right). \qquad (8\text{-}43)$$

This equation is known at the *van 't Hoff equation* after the Dutch chemist J. H. van 't Hoff. The equation was derived using the following property of log functions

$$\ln\left(\frac{K_2}{K_1}\right) = \ln K_2 - \ln K_1. \qquad (8\text{-}44)$$

Some conclusions can be drawn from the van 't Hoff equation. Consider exothermic reactions in which the standard enthalpy is negative ($\Delta_r H^\circ < 0$). In exothermic reactions, increasing the temperature decreases the equilibrium constant, which means the reaction is shifted toward less products. Decreasing the temperature in an exothermic reaction shifts the reaction toward more products. By comparison, for endothermic reactions ($\Delta_r H^\circ > 0$), increasing the temperature increases the equilibrium constant, so the reaction is shifted toward more products.

References

Barin, I, Thermochemical Data of Pure Substances, VCH Publishers, 1995.

Cropper, W.H., Great Physicists, Oxford University Press, 2001.

Feynman, R., Leighton, R.B., Sands, M., Lectures on Physics, Adeson-Wesley Publishing Company, 1966.

Pauling, L., College Chemistry, W.H. Freeman and Company, 1957.

Rosser, W.G.V., An Introduction to Statistical Physics, Ellis Horwood, 1982.

Sienko, M.J. and Plane, R.A., Chemistry, Principles and Applications, McGraw-Hill Book Company, 1979.

About the cover: Electron probability density for the hydrogen *p* orbital.

This textbook provides a set of lectures suitable for the first college chemistry course, or an equivalent self-study course at that level. The book provides historical background and insights to aid in clarifying chemistry as it is typically taught. In this text, problem sets are avoided in favor of descrption of chemical principles.

This book covers the topics:

- Atoms and molecules
- Quantum mechanics and the structure of multi-electron atoms
- Periodic behavior of atoms and the periodic table
- Chemical bonds and Lewis structures
- Kinetic theory of gases
- Thermodynamics, chemical equilibrium and entropy

www.ingramcontent.com/pod-product-compliance
Lightning Source LLC
Chambersburg PA
CBHW080917170526
45158CB00008B/2146